Sustainable Water Treatment

Innovative Technologies

Sustainable Water Treatment
Innovative Technologies

Edited by
**Zainura Zainon Noor and
Noor Salehan Mohammad Sabli**

CRC Press
Taylor & Francis Group
Boca Raton London New York

CRC Press is an imprint of the
Taylor & Francis Group, an **informa** business

CRC Press
Taylor & Francis Group
6000 Broken Sound Parkway NW, Suite 300
Boca Raton, FL 33487-2742

First issued in paperback 2019

ISBN-13: 978-1-138-03324-5 (hbk)
ISBN-13: 978-0-367-88577-9 (pbk)

Library of Congress Cataloging-in-Publication Data

Names: Noor, Zainura Zainon, editor. | Sabli, Noor Salehan Mohammad, editor.
Title: Sustainable water treatment : innovative technologies / [edited by] Zainura
    Zainon Noor; Noor Salehan Mohammad Sabli.
Description: Taylor & Francis, a CRC title, part of the Taylor & Francis imprint, a
    member of the Taylor & Francis Group, the academic division of T&F Informa,
    plc, [2017].
Identifiers: LCCN 2016037195| ISBN 9781138033245 | ISBN 9781138033252
    (eISBN)
Subjects: LCSH: Water--Purification--Technological innovations. | Sustainable
    chemistry.
Classification: LCC TD430 .S84 2017 | DDC 628.1/620286--dc23
LC record available at https://lccn.loc.gov/2016037195

Visit the Taylor & Francis Web site at
http://www.taylorandfrancis.com

and the CRC Press Web site at
http://www.crcpress.com

Contents

Section I Innovative Biological Processes for the Recovery of Value-Added Products from Wastewater

Section II MBR Technologies

Section III   Advanced Chemical-Physical Processes
for Industrial Wastewater Treatment

Preface

This book focuses on wastewater treatment with green and innovative technologies that promote sustainability. It talks about studies conducted on innovation from existing biological, chemical, and physical processes in wastewater treatment. It aims to help researchers or related parties that are interested in implementing wastewater treatment with greener technologies. Since this book covers all the fundamental processes (biological, physical, and chemical) in wastewater treatment, it will bring great benefits to readers as they would gain better understanding of green technologies in wastewater treatments.

Acknowledgments

First and foremost, all praise be to God for his blessings and guidance for giving us the inspiration to embark on this project and instilling in all of us the strength to see that this book becomes a reality. Many people have contributed to the creation and completion of this book. We would like to acknowledge with appreciation numerous valuable comments, suggestions, constructive criticisms, and praise from evaluators and reviewers. Their suggestions have greatly helped us to improve the quality of this book. We also would like to express our gratitude to all who have helped in one way or another in the planning, brainstorming, writing, and editing stages of this book especially our Green Technology Research Group. We would also like to extend our appreciation to Universiti Teknologi Malaysia for providing us with the facilities vital to the completion of this book. Finally, we would like to express our appreciation to our families for their continuous patience, understanding, and support throughout the preparation of this book.

Editors

Zainura Zainon Noor is an associate professor of chemical engineering at Universiti Teknologi Malaysia. She embarked on her career in UTM in 1999 as a research officer in a chemical engineering pilot plant prior to joining the Faculty of Chemical Engineering and Natural Resources 2 years later. A well-trained chemical engineer specializing in environmental engineering, Dr. Zainura is an intrinsically passionate individual driven toward finding greener and eco-friendly solutions. Through her unremitting interest, years of academic study, as well as conducive research and consultation activities, Dr. Zainura has established and strengthened her expertise in green technology, including cleaner production, life cycle assessment (LCA), water and carbon footprints, greenhouse gas inventory and projection as well as sustainable development. She is an accomplished project manager and is currently leading the Green Technology Research Group (Green Tech RG) at one of UTM's prominent centers of excellence, the Institute of Water and Environmental Management (IPASA). Recognizing her expertise in green technology, in 2009, the Department of Environment (DOE) Malaysia appointed her to develop the Cleaner Production Module, which was later used as the training module for the department's officers from all over Malaysia. Recently, she was selected by the Malaysian Government (under the Ministry of Natural Resources and Environment) as a consultant for the development of the Malaysia Environmental Performance Index (EPI). She is also an appointed committee member of the Green Technology Focus Group (sustainable solid waste) under the Ministry of Green Technology, Energy and Water. Dr. Zainura is also a renowned speaker and has given talks at numerous seminars, workshops, and short courses at both national and international levels. Dr. Zainura earned her PhD and MS degrees from Newcastle University, UK, and BS degree from Vanderbilt University.

Noor Salehan Mohammad Sabli is currently a PhD student in environmental engineering at the Faculty of Chemical Engineering, Universiti Teknologi Malaysia (UTM) under the supervision of an associate professor Dr. Zainura Zainon Noor. Her study is in the field of environmental engineering specializing in water footprint. Her research focuses on developing water footprint framework for calculating water usage through crude palm oil production from nursery until the mill, together with adapting the life cycle assessment approach in the framework.

Contributors

Abdul Hadi Abdullah
Department of Environmental
  Engineering
Universiti Teknologi Malaysia
Johor, Malaysia

Noor Amirah Abdul Aziz
Centre for Environmental
  Sustainability and Water
  Security
Universiti Teknologi Malaysia
Johor, Malaysia

Shreeshivadasan Chelliapan
Department of Engineering
Universiti Teknologi Malaysia
  Razak School
Kuala Lumpur, Malaysia

Mohd. Fadhil Md Din
Centre for Environmental
  Sustainability and Water
  Security
Universiti Teknologi Malaysia
Johor, Malaysia

Abdullahi Mohammed Evuti
Faculty of Engineering
University of Abuja
Abuja, Nigeria

Nor Badzilah Hasan
Department of Biosciences and
  Health Sciences
Universiti Teknologi Malaysia
Johor, Malaysia

Mohd Ariffin Abu Hassan
Department of Chemical Engineering
Universiti Teknologi Malaysia
Johor, Malaysia

Raja Kamarulzaman Raja Ibrahim
Department of Physics
Universiti Teknologi Malaysia
Johor, Malaysia

Rabialtu Sulihah Ibrahim
Faculty of Chemical and Energy
  Engineering
Universiti Teknologi Malaysia
Johor, Malaysia

Zaharah Ibrahim
Faculty of Bioscience and Medical
  Engineering
Universiti Teknologi Malaysia
Johor, Malaysia

Qistina Ahmad Kamal
Department of Biosciences and
  Health Sciences
Universiti Teknologi Malaysia
Johor, Malaysia

Siti Nurhayati Kamaruddin
Faculty of Chemical and Energy
  Engineering
Universiti Teknologi Malaysia
Johor, Malaysia

Chi Kim Lim
Faculty of Technology Management
  dan Business
Universiti Tun Hussein Onn Malaysia
Johor, Malaysia

Zaiton Abd Majid
Department of Chemistry
Universiti Teknologi Malaysia
Johor, Malaysia

Florianna Lendai Michael Mulok
Department of Human Resource
    Development
Universiti Malaysia Sarawak
Sarawak, Malaysia

Noor Sabrina Ahmad Mutamim
Universiti Malaysia Pahang
Pahang, Malaysia

Chin Hong Neoh
Centre for Environmental
    Sustainability Water Security
    (IPASA)
Universiti Teknologi Malaysia
Johor, Malaysia

Zainura Zainon Noor
Centre for Environmental
    Sustainability Water Security
    (IPASA)
Universiti Teknologi Malaysia
Johor, Malaysia

Mohanadoss Ponraj
Centre for Sustainable Technology
    and Environment (CSTEN)
Universiti Tenaga Nasional
Kajang, Malaysia

Noor Salehan Mohammad Sabli
Faculty of Chemical and Energy
    Engineering
Universiti Teknologi Malaysia
Johor, Malaysia

Venmathy Samanaseh
Faculty of Chemical and Energy
    Engineering
Universiti Teknologi Malaysia
Johor, Malaysia

Cindy Lee Ik Sing
Faculty of Chemical and Energy
    Engineering
Universiti Teknologi Malaysia
Johor, Malaysia

Tan Wei Yie
Department of Biosciences and
    Health Sciences
Universiti Teknologi Malaysia
Johor, Malaysia

Ee Ling Yong
Faculty of Civil Engineering
Universiti Teknologi Malaysia
Johor, Malaysia

Adhi Yuniarto
Department of Environmental
    Engineering
Institut Teknologi Sepuluh
    Nopember
Jawa Timur, Indonesia

Mohd Badruddin Mohd Yusof
Department of Environmental
    Engineering
Universiti Teknologi Malaysia
Johor, Malaysia

Nor Azimah Mohd Zain
Faculty Biosciences and Biomedical
    Engineering
Universiti Teknologi Malaysia
Johor, Malaysia

Section I

Innovative Biological Processes for the Recovery of Value-Added Products from Wastewater

1

Enzymatic Hydrolysis of Waste Cooking Palm Oil by PVA–Alginate–Sulfate Immobilized Lipase

Nor Badzilah Hasan, Tan Wei Yie, and Nor Azimah Mohd Zain

CONTENTS

1.1 Introduction

Fatty acids exist in nature as carboxylic acids with long hydrocarbon chains, which are either saturated or unsaturated. They consist of carbon (C), hydrogen (H), and oxygen (O) and are arranged as a carbon chain skeleton with a carboxyl group (–COOH) at one end. The hydrocarbon chain length may vary from 10 to 30 carbons but it is usually from 12 to 18 carbons. Fatty acids are usually derived from triglycerides and are the main component of vegetable oil and animal fats. Fatty acids are widely used as raw materials in food, cosmetics, the pharmaceutical and dairy industries, and skin care products. Today, the production of fatty acid and glycerol from cooking palm oil is vital especially in oleochemical industries (Serri et al., 2008). Many researchers have used enzyme-catalyzed hydrolysis in order to reduce energy consumption and minimize thermal degradation of the products. However, studies using immobilized lipase which has the ability to hydrolyze cooking palm oil into fatty acid and glycerol have not been widely explored.

Waste cooking oil (WCO) is known for its high acid value of free fatty acids (FFAs; Araujo, 1995). FFAs are value-added products because of their wide industrial applications such as soap production, surfactants manufacturing, biomedical uses, and biodiesel production (Hill, 2000; Habulin and Knez, 2002).

1.2 Waste Cooking Oil

WCO is the residue from the kitchen, restaurants, and food factories. WCOs are basically generated from vegetable oils used at high temperature in food frying. As a result, this process causes hydrolysis, polymerization, and oxidation reactions which change the physical and chemical properties of the oil.

1.2.1 Environmental Pollution Cause by WCO

Increasing production of WCO from household and industrial sources is a growing issue all around the world. Table 1.1 shows the quantity of WCO produced in selected countries. This residue usually contains large amounts of FFAs, polymers, and decomposition products besides triglyceride and

TABLE 1.1

Quantity of WCO Produced in Selected Countries

Country	Quantity (million tonnes/year)
United States	10.0
China	4.5
European Union	0.7–1.0
Japan	0.45–0.57
Malaysia	0.5
Canada	0.12
Taiwan	0.07

some diglyceride due to the reaction of oxidation and hydrogenation (Lam et al., 2010). This residue is regularly being poured down the drain resulting in a wastewater treatment problem. Besides this, the residue can be integrated into the food chain via animal feed, thus resulting in a potential human health risk (Costa Neto et al., 2000). Recently, WCO rich in fatty acids has gained great interest due to its use in biodiesel production.

1.2.2 Composition of WCO

WCOs consist of saturated and unsaturated fatty acid, for instance; waste cooking palm oil is rich in palmitic acid, oleic acid, linoleic acid, and stearic acid. The fatty acids that do not have double bonds are termed "saturated," such as stearic acid and palmitic acid. These chains contain the maximum number of possible hydrogen atoms per atom carbon. Fatty acids that have double bonds are termed "unsaturated," such as linoleic acid and oleic acid. These chains do not contain the maximum number of hydrogen atoms due to the presence of double bond(s) on some carbon atoms (Lam et al., 2010).

WCO and fatty acid compositions are summarized in Table 1.2 (Hingu et al., 2010) and Table 1.3 (Wan Omar et al., 2009). The analysis in Table 1.3 shows that

TABLE 1.2

WCO Profile

WCO Composition	Percentage
Fatty acid	32.13
Ester	42.55
Methyl ester	2.14
Ketone	2.54
Aldehyde	4.69
Alkane	2.16
Alkene	0.68
Alcohol	0.29
Other	11.41

TABLE 1.3

FFA Profile in WCO

Fatty Acid Composition	Percentage (%)
Palmitic acid C16:0	9.08
Stearic acid C18:0	2.16
Oleic acid C18:1	35.34
Linoleic acid C18:2	53.4

they are mainly composed of 90% of unsaturated fatty acids (linoleic and oleic acids) and 10% saturated fatty acids (palmitic and stearic acid).

1.3 Lipases

Lipase enzymes are important in biological systems which belong to the group of serine hydrolyses (E.C. 3.1.1.3) (Jaeger and Eggert, 2002). They are mostly found built on an alpha and beta hydrolase fold with beta sheet containing the catalytic residues. Their catalytic site is composed of serine, aspartic acid, and histidine. The interior topology of alpha and beta hydrolase fold protein is mainly composed of parallel beta pleated strands separated by an alpha helix. Figure 1.1 shows the 3D structure of lipase from *Candida rugosa*. In Figure 1.1, "a" represents the helices which are packed against the central L-sheet. "b" represents the central L-sheet while "c" represents the smaller

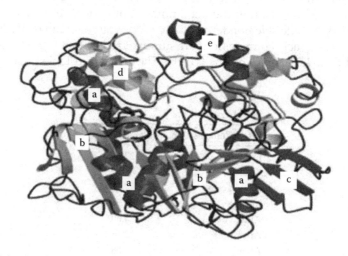

FIGURE 1.1
CRL enzyme 3D structure (Adapted from Cygler, M. and Schrag, J. D. 1999. *Biochimica et Biophysica Acta.* 1441: 205–214.).

N-terminal L-sheet. The closed conformation of the lid is represented by "d" and the open conformation is represented by "e". The residues forming the catalytic triad are indicated by "e".

Lipases are like esterases. They catalyze the hydrolysis and the transesterification of ester groups. However, the difference between esterases and lipase is that esterases act on soluble substrates while lipases catalyze reactions on water insoluble substrates and the presence of the water or lipid interface is a usual prerequisite for efficient catalysis (Cygler and Schrag, 1999). Their natural substrate is triglycerides. Lipases catalyze the hydrolysis of triglycerides to glycerol and fatty acids (Vaidya et al., 2008).

Lipases can also catalyze a wide range of enantio- and regioselective reactions such as hydrolysis, esterification, transesterification, aminolysis, and ammonolysis depending on the nature of substrate and reaction conditions (Vaidya et al., 2008). The versatility of lipase-catalyzed reactions made them greatly applied in numerous industrial processes including oils and fats, detergents, baking, cheese making, hard-surface cleaning as well as leather and paper processing (Schmidt and Verger, 1998; Jaeger et al., 1999; Villeneuve et al., 2000).

1.3.1 Lipase Sources

Various enzymes that are isolated from 2% of the world's microorganisms have been used as enzyme sources (Hasan et al., 2006). However, microbial enzymes are often more widely used compared to enzymes that are derived from plants or animals because of the great variety of catalytic activities available, high yield possibility, ease of genetic manipulation, and rapid growth of microorganisms on inexpensive media. In addition, microbial enzymes production is more convenient, safer, and exhibits higher stability (Wiseman, 1995).

In recent years, research on microbial productions of lipase has increased because of their wide application in industry such as the hydrolysis of fats, production of fatty acids, food additives, detergent additives, cosmetics, and care products (Bjorkling et al., 1991; Malcata, 1996). Lipases have been isolated from a wide number of plant, animal, and microbial sources (Sangeetha et al., 2010). Lipase producing microorganisms includes bacteria, fungi, and yeasts (Rapp, 1995). The ease of isolation of microbes' lipase has made both bacteria and fungi predominant sources of lipase. Lipases obtained from fungal sources were thought to be the best source for commercial application until bacterial lipases were discovered. Table 1.4 (Hasan et al., 2006) shows a lipase producing microorganism.

1.3.2 Lipase from *C. rugosa*

Lipases from *C. rugosa* were firstly described early in the 1960s by isolating the yeast from natural soils. It is also known due to its great lipase production (Yamada et al., 1963; Tomizuka et al., 1966). In addition, the lipases from yeast were nonpathogenic. Therefore, there were a great number of reviews about

TABLE 1.4

Isolation of Lipase from Various Microorganisms

Type of Microorganism	Lipase-Producing Microorganisms	Reference
Bacteria	*Bacillus* sp.	Imamura and Kitaura (2000)
	Bacillus subtilis	Ruiz et al. (2005)
	Pseudomonas sp.	Sarkar et al. (1998)
	Pseudomonas aeruginosa	Chartrain et al. (1993)
	Staphylococcus aureus	Gotz et al. (1998)
	Penicillium cyclopium	Chahinian et al. (2000)
	Lactobacillus plantarum	Lopes et al. (1999)
	Chromobacterium viscosum	Taipa et al. (1995)
Fungus	*Candida cylindracea*	Muralidhar et al. (2001)
	Rhizomucor miehei	Herrgard et al. (2000)
	Acinetobacter sp.	Snellman et al. (2002)
	Fusarium solani	Knight et al. (2000)

C. rugosa, highlighting its different aspects like biochemical, fermentation technology, or some biocatalytical applications (Benjamin and Pandey, 1998; Cygler and Schrag, 1999; Akoh et al., 2004). Furthermore, the lipase produced by *C. rugosa* is one of the most commonly used enzymes in organic solvents due to its high activity of hydrolysis, esterification, transesterification, and aminolysis (Villeneuve et al., 2000). The findings were also proven in the studies by Winayanuwattikun et al. (2011). In this study, *C. rugosa* lipase (CRL) also showed a broader range of substrate specificity with high activity for the substrates from 4 to 16 carbon chain length (Winayanuwattikun et al., 2011). Formally, *C. rugosa* is known as *Candida cylindracea* (Khor et al., 1985).

1.3.3 Hydrolysis Reaction of Lipase

A lipase reaction system usually consists of two immiscible phases where the water phase contains dissolved lipase while the organic phase contains dissolved substrates (triglycerides). The water phase of lipase is contacted with the triglycerides in organic phase forming liquid–liquid dispersion. Lipases catalyze the hydrolysis of triglyceride into FFAs and glycerol at the interface between the two liquids (Murty et al., 2002). Triglycerides do not dissolve in the water phase, thus the reaction has to be placed at the interface of the lipid and water phase (Pronk et al., 1988). The reaction often starts with reversible adsorption of the enzyme at the interface which then binds to the substrate to initiate catalysis (Tsai and Liching, 1990). Since the hydrolysis activity occurs at the lipid–water phase, the presence of an organic solvent is necessary to solubilize the lipid in water in order to achieve their hydrolysis catalyzed by lipases (Torres and Otero, 1996). On the other hand, emulsifiers can be added to increase the interface of the lipid–water emulsion. However,

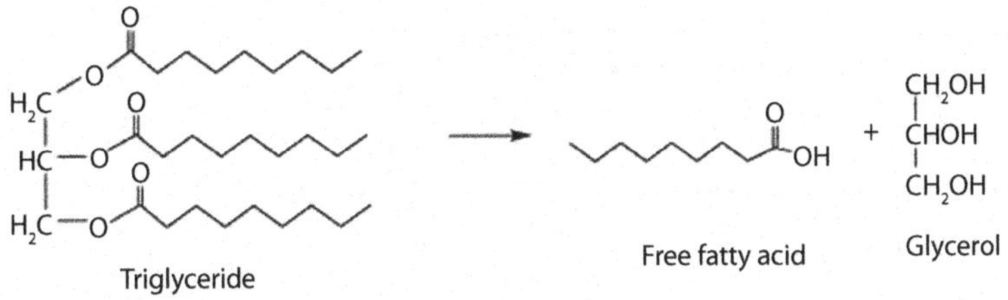

FIGURE 1.2
Hydrolysis reaction of lipase (Adapted from Gupta et al., 2008. *International Journal of Biological Macromolecules.* 42: 145–151.) The hydrolysis reaction yields 1 mole of glycerol and 3 moles of fatty acid per mole of triglycerides (Adapted from Murty, V. R., Bhat, J. and Muniswaran, P. K. A. 2002. *Biotechnology and Bioprocess Engineering.* 7: 57–66.)

according to Chew et al. (2008), the addition of emulsifiers is not preferred to avoid additional separation processes which are not favorable in industry. Figure 1.2 explains the hydrolysis reaction by lipase (Gupta et al., 2008; Villeneuve et al., 2000).

1.4 Immobilization Method

Immobilization can be defined as the technique used for physical or chemical fixation of cells, organelles, enzymes, or other proteins onto a solid support, into a solid matrix, or retained by a membrane, in order to increase their repeated or continuous use (IUPAC). There are three main types of methods used for immobilizing enzymes, such as entrapment, cross-linking, and carrier binding as shown in Figure 1.3. The selection of the immobilization method is based on several factors, such as overall enzymatic activity, cost of immobilization method, toxicity of immobilization reagents, and the effectiveness of enzyme utilization (Ozturk, 2001). Among various techniques employed for the immobilization of enzymes, entrapment methods are mostly used in cell immobilization procedures. The inert characteristics of the matrix result in relatively little damage to the native enzyme's structure. Besides that, the polymer used was inert to enzymes. Therefore, enzyme denaturation can be avoided. Furthermore, entrapped enzymes are more suitable for use with smaller size of substrates as larger enzymes will not be able to pass through the membrane and reach the active site of the biocatalyst (Villeneuve et al., 2000).

1.4.1 Advantages of Enzyme Immobilization

Enzymes are widely used as catalysts in many industrial, biomedical, and analytical processes. There has been considerable interest in enzyme

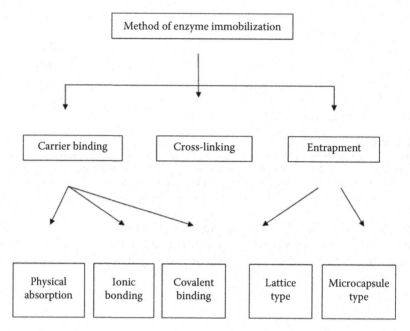

FIGURE 1.3
Methods for enzyme immobilization.

immobilization due to the simple preparation. The use of immobilized enzymes offers several advantages compared to free enzymes. The first advantage of immobilization permits the repeated use of enzymes. Although the immobilization procedure further increases the manufacturing expense, the multiple use of immobilized enzymes provides cost advantages. An ideal immobilization model is one which permits a high turnover rate of the enzyme yet retains its high catalytic activity over time. Secondly, the repeated use of enzymes allows continuous processes in industry as immobilized enzymes can be used immediately after a reaction. Thirdly, many experiments have shown that immobilized enzymes significantly enhanced stability in terms of thermal and pH aspects. Immobilization increased the thermal stability of enzymes that conferred their use for longer periods at a higher temperature compared to free enzymes. Besides that, immobilization broadened the working pH of an enzyme. This could allow enzyme activity over a wider pH (Twyman, 2005). Other than that, it enables the enzyme to be easily separated from the product. This would simplify enzyme applications; support a reliable and efficient technology, as well as provide cost advantages (Tischer and Wedekind, 1999).

1.4.2 Polyvinyl Alcohol

Polyvinyl alcohol (PVA) has been used as immobilization matrix since about 14 years ago (Hassan and Peppas, 2000). It is also the largest hydrophilic synthetic polymer produced in the world (Ramaraj, 2000). PVA is rubber elastic-like in nature, nontoxic, and odorless. PVA is a polymer of great interest due to its desirable application in the pharmaceutical and biomedical fields (Hassan and Peppas, 2000). In the biomedical field, PVA has been proposed as a promising biomaterial that is suitable for tissue mimicking, vascular cell culturing, and vascular implanting (Cygler and Schrag, 1999). For instance, PVA has been applied in tissue engineering for regenerating a wide variety of tissue and organs, including arterial phantoms, heart valves, corneal implants, and cartilage tissue substitutes.

1.4.3 PVA–Alginate as Supporting Material

Cell entrapment in polymeric matrices is widely used for cell immobilization (Wu and Wisecarver, 1992; Zhang et al., 2005). It also proves that the immobilization enzyme on polymer matrices lead to better reusability (Idris et al., 2008). The living cells are enclosed in a polymeric matrix which is porous enough to allow the diffusion of substrates into the cells and permit products of enzymatic reaction to move away from the cells (Wu and Wisecarver, 1992). A wide variety of materials have been successfully used for cell entrapment such as agar, agarose, kappa carrageenan, collagen, alginates, chitosan, polyacrylamide, polyurethane, and cellulose (Ariga et al., 1987). Recently, the use of PVA for immobilization purpose has been investigated (Wu and Wisecarver, 1992; Hashimoto and Furukawa, 1987). Ariga et al. (1987) used the freezing and thawing of the PVA method to form a gel suitable for cell immobilization. They found that this technique produced a low cost material and exhibited high strength with rubber-like elasticity (Ariga et al., 1987). PVA cross-linked with boric acid has been developed by Hashimoto and Furukawa (1987). They have used a new inexpensive and less energy-intensive immobilization method to immobilize activated sludge using PVA. They cross-linked the PVA using boric acid solution and a monodiol type of PVA–boric acid gel lattice was produced. The activated sludge was successfully immobilized in PVA–alginate beads without loss of biological activity (Hashimoto and Furukawa, 1987). A modified PVA–alginate bead was developed and reported by Idris et al. (2008) by introducing sodium sulfate. The beads produced by this technique were found to be more stable in terms of chemical and mechanical strength. The beads also displayed superior enzyme activity and showed relative good diffusivities (Zain et al., 2010). Similar findings were also reported by Takei et al. (2011).

1.4.4 Advantages of PVA–Alginate Matrix

In recent years, the use of PVA for cell immobilization has attracted wide attention due to several advantages that it offers. PVA is a cheap and nontoxic synthetic polymer as well as being easy to process. Besides this, enzymes immobilized in a PVA matrix showed high activity and high stability with repetitive use, thus increasing economic viability of biosynthetic process-ing (Zain et al., 2010). On the other hand, PVA is a hydrophilic support-ing material. Generally, the use of hydrophilic support in immobilization enhances enzyme stability while the use of hydrophobic support material appears to have its disadvantages. PVA can cause protein stabilization by its attachment to the polymer chains (Kozhukharova et al., 1988). It is suitable for use for immobilization purposes because it can be easily modified through its hydroxyl groups. PVA also offers assorted advantages over the conventional alginate matrix including lower production cost, higher robust-ness, and nontoxicity to viable cells. PVA beads exhibit rubber-like elastic-ity in nature (Hassan and Peppas, 2000). Thus, PVA beads provide stronger mechanical strength compared to alginate beads. Moreover, PVA beads display high stability within a wide range of pH that is from pH 1 to pH 13, while alginate beads are relatively stable in the range of 6–9 (Khoo and Ting, 2001). Furthermore, alginate beads encountered a weight loss up to 20% and 24% at low and high pH. By being aware of these advantages, PVA has been used widely in cell immobilization (Idris et al., 2008).

1.4.5 Drawbacks of PVA and Its Solution

The PVA–boric acid technique provides an easy and low cost method in enzyme immobilization. However, there are some problems when using the PVA–boric acid method for immobilization. PVA is a sticky material, thus PVA beads have the tendency to agglomerate (Wu and Wisecarver, 1992). Nevertheless, this matrix is still used by many researchers. In order to eliminate the agglomeration, calcium alginate has been introduced. It also serves to improve the surface properties of the beads (Wu and Wisecarver, 1992; Yujian et al., 2006). The application of calcium alginate in a mixture with PVA for enzyme immobilization has been reported by Wu and Wisecarver (1992). The PVA–alginates beads produced were proven to be very strong and durable with no biological loss for 2 weeks of continuous operation in a fluidized bed reactor.

The introduction of sodium alginate in the PVA–boric acid method was also suggested by Slokoska et al. (1999). The finding of Slokoska et al. (1999) demonstrated that the photo-cross-linked PVA and calcium alginates beads are suitable for the entrapment of fungal cells (Slokoska et al., 1999). PVA exhibits a high degree of swelling in water (Hassan and Peppas, 2000). It will readily dissolve in aqueous solution causing the enzyme to leak out from the matrix (Zain et al., 2011). Therefore, the PVA must cross-link either chemically or physically to make it soluble. The most popular cross-linking reagent for immobilization is glutaraldehyde (Villeneuve et al., 2000). Besides that, cross-linking the PVA using boric acid solution to produce a monodiol type PVA–boric acid gel lattice has also been reported by Hashimoto and Furukawa and Wu and Wisecarver. The other drawback of this polymer is that the saturated boric acid solution is highly acidic (pH < 4) and it causes a drastic decrease in the viability of immobilized cells. This obstacle can be overcome by adding sodium sulfate which acts as an inducer for cross-linkage of PVA to avoid the drastic decrease in cell viability caused by saturated boric acid solution (Idris et al., 2008; Takei et al., 2011).

1.5 Case Study

1.5.1 Materials

CRL (3.1.1.3) (Type 1176 U/mg) was purchased from Sigma Aldrich (Japan). PVA 60,000 MW and boric acid were purchased from Merck Schuchardt OHG, Darmstadt, Germany. Sodium alginate was obtained from FlukaChemie GmbH, Buchs, sodium sulfate from GCE Laboratory Chemicals, and calcium chloride from R&M Marketing, Essex, UK. Iso-octane with 99.84% assay was purchased from Fisher Chemicals (UK). Other reagents used were analytical reagent grades and used without further purification including phosphate buffer solution pH 7.5 or otherwise stated.

1.5.2 Pretreatment of WCO

WCO was obtained from a food stall near Universiti Teknologi Malaysia (UTM). For a successful reaction, the oil must be free from water and other impurities. Initially the samples of waste cooking palm oil were filtered to remove any suspended food particles. Then, the waste cooking palm oil was heated at 105°C for 1 h to remove its water content. After that, the titrimetry method with NaOH was used to determine the FFA content in the WCO (Patil et al., 2010).

1.5.3 Preparation of Lipase Enzyme Solution

The pH of the phosphate buffer solution was adjusted to pH 7.5. CRL (5 g) was dissolved in 100 mL of phosphate buffer solution. Then, the enzyme solution was filtered using a 0.45 μm nylon syringe filter to sterilize the enzyme. The sterilized enzyme solution was stored at 4°C until further used.

1.5.4 Immobilization of CRL

CRL solution with the volume of 10 mL, 5% (v/v) was mixed with 90 mL PVA–alginate solution. The mixture was mixed comprehensively and introduced as drops by using a rotary pump into a 100 mL mixed solution of saturated boric acid 5% (w/v) and calcium chloride 2% (w/v). The beads were stirred gently for 30–50 min to complete the solidification. Then, the PVA–alginate beads were stored at 4°C for 24 h. After 24 h, the mix solution of boric acid and calcium chloride was discarded and replaced with a 7% (v/v) boric acid solution. The beads were stirred in the boric acid solution for 30 min and the solution was then replaced with 0.5 M of sodium sulfate solution and stirred for another 30 min. Then, the beads were kept at 4°C until further used (Idris et al., 2008).

1.5.5 Waste Cooking Palm Oil Hydrolysis

A conical flask of 250 mL was initially filled with 3 g of cooking palm oil and 30 mL of iso-octane solvent. Phosphate buffer solution (30 mL, pH 7.0; unless otherwise stated) was added into the conical flask so that the ratio of oil to aqueous (buffer solution) is 1. The mixture formed two layers. Three other identical mixtures as above were prepared. To start the reaction, 0.3 g of CRL was added to three flasks of reaction mixtures and one was left without the CRL for control measurement. The mixtures were agitated in the orbital shaker at 45°C at 200 rpm. Samples were withdrawn from the oil every 30 min. The same procedure was carried out using immobilized lipase (Serri et al., 2008). To determine the effect of parameters on the hydrolysis of WCO, three variables were taken into consideration which is pH, temperature, and enzyme concentration. The pH varied from 7 to 8. The temperature varied from 30°C to 50°C and enzyme loading varied from 2 to 8 g of beads which correspond to 96.43 to 385.73 U/mL enzyme.

1.5.6 Determination of Degree of Hydrolysis (Conversion) and Rate of Hydrolysis

The degree of hydrolysis was determined by titration of the oil phase samples with 0.1 M sodium hydroxide (NaOH). To each sample, 5 mL of the oil phase was dissolved in 5 mL ethanol:diethyl ether (1:1% v/v). The amount of 0.1 M NaOH required to neutralize the acid was noted. A blank titration was done as a control sample. Phenolphthalein was used as an indicator. The degree of hydrolysis, X is calculated as below (Serri et al., 2008).

$$X\% = \frac{\text{(ml NaOH used) (molarity of NaOH)} \text{ (average molecular weight of fatty acid)}}{10 \text{ (weight of sample)}} \tag{1.1}$$

1.5.7 Fatty Acid Concentration Determination

All samples collected were analyzed using Perkin Elmer Autosystem XL equipped with a flame-ionization and a Nukol™ 15 m × 0.53 mm i.d. column coated with 0.5 µm (25326) column. Helium served as a carrier gas at a flow rate of 20 mL/min. The column temperature was programmed from 110°C to 220°C with the increase of 8°C/min. The injector and detector were monitored at 250°C and the amount of sample injection was 0.2 μL with direct injection. The presence of fatty acid was based on the comparison of retention time and peak area of the sample with oleic acid as standard.

1.6 Results and Discussions

1.6.1 Effects of Temperature on Immobilized CRL

The reaction temperature is an important parameter in enzyme catalysis. The effect of temperature effect is significant because in order to increase the reaction rate, the temperature has also to be increased. This fact is also supported by Kumari et al. (2009) where an increase in temperature will speed up enzyme-mediated reaction. In this study, the reaction mixture was incubated at temperatures varying in the range of 30–50°C with the immobilized lipase. As shown in Figure 1.4, the highest fatty acid concentration and hydrolysis conversion were achieved at 50°C.

If the enzyme is immobilized on a suitable support, thermal stability is usually improved. Therefore, determination of the optimum temperature at which the lipase does not lose its activity is very important in finding the optimum operating conditions for immobilization. Previous study by Dave and Madamwar (2006) also reported that the optimum temperature was found at 50°C when *C. rugosa* lipase was immobilized in PVA–alginate matrix for the esterification reaction. Similarly, an optimum temperature of 50°C was reported by Santos et al. (2008) when using *C. rugosa* lipase immobilized by covalent attached on polysiloxane–polyvinyl alcohol for the hydrolysis of olive oil.

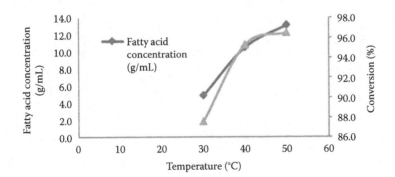

FIGURE 1.4
Effect of temperature on fatty acid production and hydrolysis conversion by using immobilized CRL (pH = 7.0, enzyme loading = 8 g of immobilized beads [385.73 U/mL], 200 rpm).

1.6.2 Effects of pH on Immobilized CRL

As enzyme activity also changes with pH, choice of the working pH also depends on the optimum working pH of the enzyme. Therefore, in order to maximize the immobilization yield, to work in a suitable pH range is essential. The effect of pH of the reaction medium on hydrolytic activity of the immobilized lipase was evaluated by adjusting the pH in the range of 7–8 at 50°C. A pH of 7 for the immobilized lipase was found to be optimum for achieving efficient hydrolysis with highest fatty acid production and highest hydrolysis conversion (Figure 1.5).

Similarly to this study, an optimum pH of 7 was reported by Garcia et al. (1992) when a lipase from *C. rugosa* was immobilized by adsorption on flat sheets made of microporous polypropylene for the hydrolysis of milk fat triglycerides. Kang and Rhee also obtained an optimum pH of 7 when using *C. rugosa* lipase immobilized by adsorption on swollen Sephadex for the hydrolysis of olive oil. However, Santos et al. obtained an optimum pH of 8 when a lipase from *C. rugosa* was immobilized on poly(N-methylolacrylamide) by physical adsorption. Lipases undergo structural changes in some pH values and this leads to inactivation of the enzyme or change in its activity due to perturbation in the vicinity of the active site. Similarly, these reactions might have occurred in CRL proteins, causing low activity at pH > 8.0 (Akova and Üstün, 2000). Therefore, the working pH depends mainly on the method of immobilization and the interaction between enzyme and support (Ting et al., 2006).

1.6.3 Effect of Enzyme Loading on Immobilized CRL

The effect of enzyme loading on the hydrolysis reaction was also investigated. The enzyme loading varied from 96.43 to 385.73 U/mL which corresponds to 2–8 g. Figure 1.6 shows that the maximal fatty acid production increases with the biocatalyst loading.

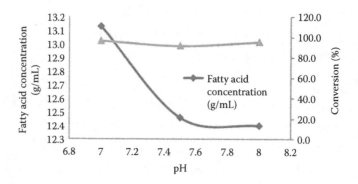

FIGURE 1.5

Effect of pH on fatty acid production and hydrolysis conversion by using immobilized CRL (temperature = 50°C, enzyme loading = 8 g of immobilized beads (385.73 U/mL); 200 rpm).

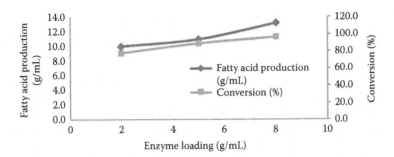

FIGURE 1.6
Effect of enzyme loading on fatty acid production and hydrolysis conversion by using immobilized CRL (temperature = 50°C, enzyme loading = 8 g of immobilized beads, 385.73 U/mL; 200 rpm).

The optimum enzyme loading was found to be 385.73 U/mL which is equal to 8 g of immobilized beads. A study by Akova and Üstün (2000) suggested that at low loadings, there is a large excess of surface area that the enzyme can occupy and the lipase attempts to maximize its contact with the surface, which results in a loss of conformation and consequently in a reduction of activity. As the loading increases, less area is available for the lipase to spread itself, more of its active conformation is retained, and the loss in activity is reduced. However, in the presence of an excessive amount of lipase, the active site of the enzyme cannot be exposed to the substrate and many molecules in the enzyme tend to aggregate together (Liou et al., 1998; Foresti and Ferreira, 2005). On the other hand, Salis et al. (2008) obtained the highest enzyme activity which was 600 mg/g (8.35 kLU/g) when CRL was immobilized on macroporous polypropylene. It is suggested that the optimum enzyme loading needed for reaction also depends on the interaction between support and enzyme.

1.6.4 Comparative Study of Free and Immobilized CRL

The maximum yield of fatty acid was 13.13 g/L (of which 96.5% hydrolysis conversion was achieved) with 8 g of immobilized beads (385.73 U/mL), working pH 7 at 50°C. The same condition was also performed on free enzymes to study the hydrolytic activity. The results obtained were compared with immobilized CRL.

Based on Figures 1.7 and 1.8, PVA–alginate immobilized CRL showed the highest production of fatty acid and conversion of hydrolysis with 13.13 g/mL and 96.5%, respectively. While that, the production of fatty acid and hydrolysis conversion for free lipase was 2.9 g/mL and 68.75%, respectively. Immobilized enzymes offer a lot of advantages ranging from increasing enzyme activity to withstanding environmental stress. Immobilization reduces contamination risks, allows enzyme reuse, increases stability, and rapidly gives positive results compared to free enzymes (Idris et al., 2008; Zain et al., 2011). According

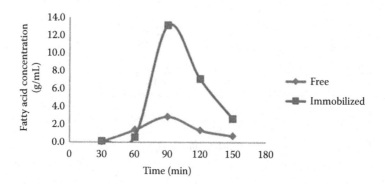

FIGURE 1.7
The fatty acid production for WCO by using free and immobilized enzyme (temperature = 50°C; pH = 7.0; enzyme loading = 8 g of immobilized beads, 385.73 U/mL; 200 rpm).

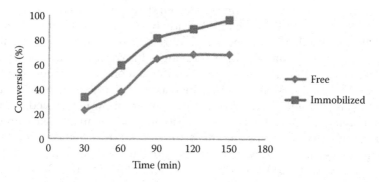

FIGURE 1.8
The hydrolysis conversion for WCO by using free and immobilized enzyme (temperature = 50°C; pH = 7.0; enzyme loading = 8 g of immobilized beads, 385.73 U/mL; 200 rpm).

to Saraiva Silva et al. (2004) PVA/alginate immobilized lipase appears to be more positive to hydrolysis of oil compared to free lipase due to the better prevailing interface condition between the PVA and the alginate. Another study was conducted by Shah and Gupta (2007) whereby the immobilized lipase gave higher biodiesel production compared to the free enzyme as it provides larger surface area of the biocatalyst preparation. Besides, the powder form of the enzyme tends to aggregate in low water media creating problems in mass transfer resulting in the low activity of lipase.

1.7 Conclusion

The enzyme acts as a biocatalyst and it is much more expensive than other catalysts. This is the main reason for using PVA–alginate beads to immobilize lipase. By immobilizing the enzyme, it is expected that it could be reused several times. Immobilization also helps in enhancing enzyme catalytic activity, enzyme stability, reusability, and recovery of enzymes. Based on the case study, immobilized lipase successfully produced more fatty acid and hydrolysis conversion compared to free lipase. This is the first time such modified PVA–alginate matrix is used to immobilize lipase to treat WCO. Lipase from *C. rugosa* was immobilized in modified PVA–alginate matrix using entrapment and the cross-linking method. PVA–alginate beads were chosen because they are a nontoxic synthetic polymer and also low cost.

References

Akoh, C. C., Lee, G. C., and Shaw, J. F. 2004. Protein engineering and applications of *Candida rugosa* lipase isoforms. *Lipids*. 39(6): 513–526.

Akova, A. and Üstün, G. 2000. Activity and adsorption of lipase from *Nigella sativa* seeds on celite at different pH values. *Biotechnology Letters*. 22: 355–359.

Araujo, J. (1995). Oxidacao de Lipidios. In: University Press (Ed.). *Quimica Dealimentos, Teoria e Pratica* (pp. 1–64). Viçosa: Universidad Federal de Viçosa.

Ariga, O., Takag, H., Nishizawa, H., and Sano, Y. 1987. Immobilization of micro-organisms with PVA hardened by iterative freezing and thawing. *Journal of Fermentation Technology*. 65(6): 651–658.

Benjamin, S. and Pandey, A. 1998. *Candida rugosa* lipases: Molecular biology and versatility in biotechnology. *Yeast*. 14: 1069–1087.

Bjorkling, F., Godtfredsen, S. E., and Kirk, O. 1991. The future impact of industrial lipases. *Trends Biotechnology*. 9: 360–363.

Chahinian, H., Vanot, G., Ibrik, A., Rugani, N., Sarda, L., and Comeau, L. C. 2000. Production of extracellular lipases by *Penicillium cyclopium* purification and characterization of a partial acylglycerol lipase. *Biosciences Biotechnology Biochemical*. 64: 215–222.

Chartrain, M., Katz, L., Marcin, C., Thien, M., Smith, S., and Fisher, E. 1993. Purification and characterization of a novel bioconverting lipase from *Pseudomonas aeruginosa* MB 5001. *Enzyme and Microbial Technology*. 15: 575–580.

Chew, Y. H., Chua, L. S., Cheng, K. K., Sarmidi, M. R., and Abdul Aziz, R. 2008. Kinetic study on the hydrolysis of palm Olein using immobilized lipase. *Biochemical Engineering Journal*. 39: 516–520.

Costa Neto, P. R., Rossi, L., Zagonel, G., and Ramos, L. 2000. Production of Biofuel Alternative to Diesel Oil Through the Transesterification of Fried Used Soybean Oil. *Quimica Nova*. 23: 531–537.

Cygler, M. and Schrag, J. D. 1999. Review structure and conformational flexibility of *Candida rugosa* lipase. *Biochimica et Biophysica Acta*. 1441: 205–214.

Dave, R. and Madamwar, D. 2006. Esterification in organic solvents by lipase immobilized in polymer of PVA–alginate–boric acid. *Process Biochemistry*. 41: 951–955.

Foresti, M. L. and Ferreira, M. L. 2005. Solvent-free ethyl oleate synthesis mediated by lipase from *Candida antarctica* B absorbed on polypropylene powder. *Catalysis Today*. 107–108: 23–30.

Garcia, H. S., Malcata, F. X., Hill, C. G., and Amundson, C. H. 1992. Use of *Candida rugosa* lipase immobilized in a spiral wound membrane reactor for the hydrolysis of milk fat. *Enzyme Microbial Technology*. 14(7): 535–545.

Gotz, F., Verheij, H. M., and Rosenstein, R. 1998. *Staphylococcal lipases*: Molecular characterisation, secretion, and processing. *Chemistry and Physical Lipids*. 93(1–2): 15–25.

Gupta, S., Yogesh, Javiya, S., Bhambi, M., Pundir, C. S., Singh, K., and Bhattacharya, A. 2008. Comparative study of performances of lipase immobilized asymmetric polysulfone and polyether sulfone membranes in olive oil hydrolysis. *International Journal of Biological Macromolecules*. 42: 145–151.

Habulin, M. and Knez, Z. 2002. High-pressure enzymatic hydrolysis of oil. *European Journal of Lipid Science and Technology*. 104: 381–386.

Hasan, F., Shah, A. A., and Hameed, A. 2006. Industrial applications of microbial lipases. *Enzyme and Microbial Technology*. 39: 235–251.

Hassan, C. M. and Peppas, N. A. 2000. Cellular PVA hydrogels produced by freeze/thawing. *Journal of Applied Polymer Science*. 76: 2075–2079.

Hashimoto, S. and Furukawa, K. 1987. Immobilization of activated sludge by PVA–boric acid method. *Biotechnology Bioengineering*. 30: 52–59.

Herrgard, S., Gibas, C. J., and Subramaniam, S. 2000. Role of electrostatic network of residues in the enzymatic action of *Rhizomucor miehei* lipase family. *Biochemistry*. 39: 2921–2930.

Hill, K. 2000. Fats and oils as oleochemical raw materials. *Pure and Applied Chemistry*. 72: 1255–1264.

Hingu, S. M., Gogate, P. R., and Rathod, V. K. 2010. Synthesis of biodiesel from waste cooking oil using sonochemical reactors. *Ultrasonics Sonochemistry*. 17: 827–832.

Idris, A., Mohd Zain, N. A., and Suhaim, M. S. 2008. Immobilization of Baker's yeast invertase in PVA–alginate matrix using innovative immobilization technique. *Process Biochemistry*. 4: 3331–3338.

Imamura, S. and Kitaura, S. 2000. Purification and characterization of a monoacylglycserol lipase from the moderately thermophilic *Bacillus* sp. H-257. *Journal of Biochemistry*. 127: 419–425.

Jaeger, K. E., Dijkstra, B.W., and Reetz, M. T. 1999. Bacterial biocatalysts: Molecular biology, three-dimensional structures, and biotechnological applications of lipases. *Annual Review of Microbiology*. 53: 315–351.

Jaeger, K. E. and Eggert, T. 2002. Lipases for biotechnology. *Current Opinion in Biotechnology*. 13: 390–397.

Khoo, K. M. and Ting, Y. P. 2001. Biosorption of gold by immobilized fungal biomass. *Biochemical Engineering Journal*. 8: 51–59.

Khor, H. T., Tan, N. H., and Chua, C. L. 1985. Lipase-catalyzed hydrolysis of palm oil. *Journal of American Oil Chemistry Society*. 63(4): 538–540.

Lam, M. K., Lee, K. T., and Mohamed, A. R. 2010. Homogeneous, heterogeneous and enzymatic catalysis for transesterification of high free fatty acid oil (waste cooking oil) to biodiesel: A review. *Biotechnology Advances*. 28: 500–518.

Liou, Y. C., Marangoni, A. G., and Yada, R. Y. 1998. Aggregation behaviour of *Candida rugosa* lipase. *Food Research International*. 31: 243–248.

Lopes, M. F. S., Cunha, A. E., Clemente, J. J., Carrondo, M. J. T., and Crespo, M. T. B. 1999. Influence of environmental factors on lipase production by *Lactobacillus plantarum* . *Applied Microbiology and Biotechnology*. 51: 249–254.

Knight, K., Carmo, M., Pimentel, B., Morais, M. M. C., Ledingham, W. M., and Filho, J. L. L. 2000. Immobilization of lipase from *Fusarium solani* FS1. *Brazilian Journal of Microbiology*. 31: 220–222.

Kozhukharova, A., Kirova, N., Popova, Y., Batsalova, K., and Kunchev, K. 1988. Properties of glucose oxidase immobilized in gel of polyvinylalcohol. *Biotechnology and Bioengineering*. 32: 245–248.

Kumari, A., Mahapatra, P., Garlapati, V.K., and Banerjee, K. 2009. Enzymatic trans-esterification of jatropha oil. *Biotechnology for Biofuels*. 2: 1.

Malcata, F. X. 1996. Engineering of/with lipases: Scope and strategies. In: Malcata, F. X. (Ed.). *Engineering of/with Lipases* (pp. 1–16). Dordrecht, Netherlands: Kluwer Academic Publishers.

Muralidhar, R. V., Chirumamilla, R. R., Ramachandran, V. N., Marchan, T. R., and Nigam, P. 2001. Racemic resolution of RS-baclofen using lipase from *Candida cylindracea*. *Meded Rijksuniv Gent Fak Landbouwkd Toegep Biol Wet*. 66: 227–232.

Murty, V. R., Bhat, J., and Muniswaran, P. K. A. 2002. Hydrolysis of oils by using immobilized lipase enzyme: A review. *Biotechnology and Bioprocess Engineering*. 7: 57–66.

Ozturk, B. 2001. *Immobilization of Lipase from Candida Rugosa on Hydrophobic and Hydrophilic Supports. Master.* Turkey: İzmir Institute of Technology.

Patil, P., Deng, S., Isaac Rhodes, J., and Lammers, R. J. 2010. Conversion of waste cooking oil to biodiesel using ferric sulfate and supercritical methanol processes. *Fuel*. 89(2): 360–364.

Pronk, W., Kerkhof, P. J. A., Helden, C., and Reit, K. V. 1988. The hydrolysis of triglycerides by immobilized lipase in a hydrophilic membrane reactor. *Biotechnology and Bioengineering*. 32: 512–518.

Ramaraj, B. 2000. Crosslinked poly(vinyl alcohol) and starch composite films. II. Physicomechanical, thermal properties and swelling studies. *Journal of Applied Polymer Science*. 103(2): 909–916.

Rapp, P. 1995. Production, regulation, and some properties of lipase activity from *Fusarium oxysporum* f.sp. *vasinfectum*. *Enzyme and Microbial Technology*. 17: 832–838.

Ruiz, C., Pastor, F. I., and Diaz, P. 2005. Isolation of lipid- and polysaccharide-degrading microorganisms from subtropical forest soil, and analysis of lipolytic strain *Bacillus* sp. CR-179. *Letters in Applied Microbiology*. 40: 218–227.

Salis, A., Pinna, M., Monduzzi, M., and Solinas, V. 2008. Comparison among immobilized lipases on macroporous poly-propylene towards biodiesel synthesis. *Journal of Molecular Catalysis B: Enzymatic*. 54: 19–24.

Sangeetha, R., Geetha, A., and Arulpandi, I. (2010). Concomitant and production of protease and lipase by *Bacillus licheniformis* VSG1: Production, purification and characterization. *Brazilian Journal of Microbial*. 41: 179–185.

Santos, J. C., Mijone, P. D., Nunes, G. F. M., Perez, V. H., and de Castro, H. F. 2008. Covalent attachment of *Candida rugosa* lipase on chemically modified hybrid matrix of polysiloxane–polyvinyl alcohol with different activating compounds. *Colloids and Surfaces B: Biointerfaces*. 61: 229–236.

Saraiva Silva, G., Fernandez, L. R. V., Higa, O. Z., Vítolo, M., and De Queiroz, M. A. A. A. 2004. Alginate-Poly(Vinyl Alcohol) Core–Shell Microspheres For Lipase Immobilization. *XVI Congressor Brasileiro de Engenharia e Ciencia dos Materiais. Porto Alegre,-RS de 28 de novembro a 02 de dezembro de 2004.*

Sarkar, S., Sreekanth, B., Kant, S., Banerjee, R., and Bhattacharyya, B. C. 1998. Production and optimization of microbial lipase. *Bioprocess Engineering.* 19: 29–32.

Schmidt, R. D. and Verger R. 1998. Lipases: Interfacial enzymes with attractive applications. *Angewandte Chemie International Edition English.* 37: 1608–1633.

Serri, N. A., Kamarudin, A. H., and Abdul Rahaman, S. N. 2008. Preliminary studies for production of fatty acids from hydrolysis of cooking palm oil using *C. rugosa* lipase. *Journal of Physical Science.* 19(1): 79–88.

Shah, S. and Gupta, M. N. 2007. Lipase catalyzed preparation of biodiesel from *Jatropha* oil in a solvent free system. *Process Biochemistry.* 42(2): 409–414.

Slokoska, L., Angelova, M., Pashova, S., Petricheva, E., and Konstantinov, C. 1999. Production of acid proteinase by *Humicola lutea* 120–5 immobilized in mixed photo-cross-linked polyvinyl alcohol and calcium-alginate beads. *Process Biochemistry.* 34: 73–76.

Snellman, E. A., Sullivan, E. R., and Colwell, R. R. 2002. Purification and properties of the extracellular lipase, LipA, of *Acinetobacter* sp. RAG-1. *European Journal of Biochemistry.* 269: 5771–5779.

Taipa, M. A., Aires-Barros, M. R., and Cabral, J. M. S. 1995. Purification of lipases. *Journal of Biotechnology.* 26: 111–142.

Takei, T., Ikeda, K., Ijima, H., and Kawakami, K. 2011. Fabrication of poly(vinyl alcohol) hydrogel beads crosslinked using sodium sulfate for microorganism immobilization. *Process Biochemistry.* 46: 566–571.

Ting, W. J., Tung, K. Y., Giridhar, R., and Wu, W. T. 2006. Application of binary immobilized *Candida rugosa* lipase for hydrolysis of soybean oil. *Journal of Molecular Catalysis B: Enzymatic.* 42(1–2): 32–38.

Tischer, W. and Wedekind, F. 1999. Immobilized enzymes: Methods and applications. *Topics in Current Chemistry.* 200: 95–126.

Tomizuka, N., Ota, Y., and Yamada, K. 1966. Studies on lipase from *Candida cylindracea*: Part I. Purification and properties. *Agricultural and Biological Chemistry.* 30(6): 576–584.

Torres, C. and Otero, C. 1996. Influence of the organic solvents on the activity in water and the conformation of *Candida rugosa* lipase: Description of a lipase-activating pretreatment. *Enzyme and Microbial Technology.* 19: 594–600.

Tsai, S. W. and Liching, C. 1990. Kinetics, mechanism, and time course analysis of lipase-catalyzed hydrolysis of high concentration olive oil in AOT-isooctane reversed micelles. *Biotechnology Bioengineering.* 38: 206–211.

Twyman, R. M. 2005. Enzymes: Immobilized enzymes. In: Worsfold, P., Townshend, A., and Poole, C. (eds.) *Encyclopedia of Analytical Science* (2nd ed). London: Elsevier Science, pp. 523–529.

Vaidya, B. K., Ingavle, G. C., Ponrathnam, S., Kulkarni, B. D., and Nene, S. N. 2008. Immobilization of *Candida rugosa* lipase on poly(allyl glycidylether-co-ethylene glycol dimethacrylate) macroporous polymer particles. *Bioresource Technology.* 99: 3623–3629.

Villeneuve, P., Muderhwa, J. M., Graille, J., and Hass, M. J. 2000. Customizing lipases for biocatalysis: A survey of chemical, physical and molecular biological approaches. *Journal of Molecular Catalysis B–Enzyme.* 9: 113–148.

Wan Omar, W. N. N., Nordin, N., Mohamed, M., and Amin, N. A. S. 2009. A two-step biodiesel production from waste cooking oil: Optimization of pre-treatment step. *Journal of Applied Sciences.* 9: 3098–3103.

Wu, K.Y. and Wisecarver, K. D. 1992. Cell immobilization using PVA crosslinked with boric acid. *Biotechnology and Bioengineering.* 29: 447–449.

Winayanuwattikun, P., Kaewpiboon, C., Piriyakananon, K., Chulalaksananukul, W., Yongvanich, T., and Svasti, J. 2011. Immobilized lipase from potential lipolytic microbes for catalyzing biodiesel production using palm oil as feedstock. *African Journal of Biotechnology.* 10(9): 1666–1673.

Wiseman, A. 1995. Introduction to Principles. In: Wiseman A. (Ed.). *Hand-Book of Enzyme Biotechnology* (3rd ed). Padstow, Cornwall, UK: Ellis Horwood Ltd. T.J. Press Ltd. pp. 3–8.

Yamada, K., Machida, H., Higashi, T., Koide, A., and Ueda, K. 1963. Studies on the production of lipase by microorganisms. *Journal of Agriculture Chemistry Society Japan.* 37: 645.

Yujian, W., Xiaojuan, Y., Hongyu, L., and Wei, T. 2006. Immobilization of acidithio-*Bacillus ferrooxidans* with complex of PVA and sodium alginate. *Polymer Degradation and Stability.* 91: 2408–2414.

Zain, N. A., Suhaimi, M. S., and Idris, A. 2010. Hydrolysis of liquid pineapple waste by invertase immobilized in PVA–alginate Matrix. *Biochemical Engineering Journal.* 50: 83–89.

Zain, N. A., Suhaimi, M. S., and Idris, A. 2011. Development and modification of PVA–alginate as a suitable immobilization matrix. *Process Biochemistry.* 46: 2122–2129.

Zhang, Y. M., Rittmann, B. E., and Wang, J. L. 2005. High-carbohydrate wastewater treatment by IAL-CHS with immobilized *Candida tropicalis. Journal of Process Biochemistry.* 40: 857–863.

2

Bioremediation of Palm Oil Mill Effluent for Itaconic Acid Production by Aspergillus terreus NRRL 1960 Immobilized in PVA–Alginate–Sulfate Beads

Qistina Ahmad Kamal and Nor Azimah Mohd Zain

CONTENTS

2.1 Introduction

Palm oil mill effluent (POME) is recognized as waste that could pollute the environment. This is due to its high value of chemical oxygen demand (COD), biological oxygen demand (BOD), suspended solids, total solids, ammoniacal nitrogen, total nitrogen, oil and grease, and color. It is also regarded as unworthy waste. The raw POME contains a very high concentration of proteins, carbohydrates, nitrogenous compounds, lipid, and minerals that lead to the possibilities of reuse as fermentation media. It can be regarded as an unpolished germ. Much effort has been put to explore the potential of POME to create value-added products. In this study, POME was used to produce itaconic acid. Itaconic acid fermentation from glucose and sucrose has a problem of high substrate cost and a very low yield. If utilization of the raw POME to produce itaconic acid is possible, it will reduce the cost to produce useful itaconic acid and thus lower the price. In addition, the immobilization process is predicted to increase the yield of this product since immobilization lowers the growth rate and relatively increases itaconic acid synthesis.

2.2 The Malaysian Palm Oil Industry

The palm oil industry started about 100 years ago when the first oil palm estate was established in 1911. That was the time when the palm plant was commercially exploited as an oil crop (Basiron and Weng, 2004). In the 1950, the palm oil industry expanded due to a change in the government's policy that diverted the main agricultural interest from rubber to oil palm. This cautious policy was implemented with intention to raise the socio-economic status of the Malaysian people. Ever since 1971, Malaysia has replaced Nigeria as the main producer of palm oil (Dalimin, 1995). Today, Malaysia has become the world's largest producer and also an exporter of palm oil resulting from the growth of Malaysia's palm oil industry over the years.

The palm oil industry has become the economic backbone of Malaysia and continuously faces challenges along its way. The Malaysian palm oil industry has grown rapidly and hence become a strong industry with an agricultural base, with a production of crude palm oil in the year 2013 of 19.21 million tonnes (Choo Yuen May, 2014). The high production of crude palm oil has made the industry an important contributor to Malaysia's GDP. The government income from palm kernel oil, palm oil, and its related products in 1998 reached almost US 5.6 billion dollars, which is equivalent to 5.6% of the GDP (Yusoff, 2006). The palm oil industry has contributed to the country income in terms of foreign exchange and increased the standard of living for the Malaysian people (Wu et al., 2007). The industry also provides income to poor families and individual holders attached to government schemes. In addition, it also provides job opportunities to agricultural workers (Khalid and Wan Mustafa, 1992; Ma et al., 1993).

According to Prasertsan and Prasertsan (1996) the milling and processing sector development processes are closely related to oil palm planting. The wet process has been implemented for palm oil milling in Malaysia because the dry process used in south Thailand was proved to be unsuitable for large-scale production. Since 1960, palm oil mills have increased greatly in number from 10 mills (Ma et al., 1993) to 410 operated mills in 2008 (MPOB, 2008). The growth of this sector fulfills the demands of crude palm oil locally and internationally.

Nevertheless, the high production of crude palm oil has resulted in a very large amount of POME being produced in Malaysia with at least 30 million tonnes of POME in 2012 (Tengku et al., 2012). It is estimated that as the demand for palm oil increases, the amount of waste generated from the process will increase. With this disturbing figure, the Malaysian palm oil mill industry has been identified as one sector that generates the highest amount of pollution load to rivers throughout the country (Hwang et al., 1978). In Malaysia, palm oil is utilized in the production of palm oil diesel or methyl ester for cars and bus usage, and major expansion of the palm oil diesel is expected to be 5% (Kalam and Masjuki, 2002; Reijnders and Huijbregts, 2008).

2.3 Palm Oil Mill Effluent

Production of palm oil has generated waste known as POME. It is an effluent produced during palm oil production and has the characteristics of being nontoxic in nature, predominantly organic, and a viscous acidic brown liquid with highly unpleasant odor according to Kathiravale and Ripin (1997). According to Ahmad et al. (2003), it is estimated that 5–7.5 tonnes of water are required in the process of producing each ton of crude palm oil. Hence, 50% or more of this water ends up as POME. With annual production figures in 2000 and 2008 as 8.3 and 16.3 million tonnes, respectively, Malaysia has become one of the largest producers and suppliers of palm oil in the world, (Foo and Hameed, 2010) with total of 423 mills processing up to 89 million tonnes of oil palm fresh fruit bunches per year. The generated POME annually is about 66.8 million tonnes according to Vairappan and Yen (2008). For every 100 tonnes of fresh fruit bunches processed, it was estimated that, 78 tonnes will end up as waste (Vikineswary et al., 1997).

Raw POME is a colloidal suspension that consists of 95%–96% water, 0.6%–0.7% oil, and 4%–5% total solids. About 2%–4% of suspended solids contribute to the total solids, which are mainly constituted of debris from palm fruit mesocarp generated from three main sources, firstly from the sterilizer condensate, secondly from separator sludge, and lastly from hydro cyclone wastewater (Borja and Banks, 1994; Khalid and Wan Mustafa, 1992; Ma, 2000). If the effluent is discharged into rivers without any treatment, it will certainly cause detrimental effects to the environment (Davis and Reilly, 1980). This is due to its high biochemical oxygen demand (25,000 mg/L), COD (53,630 mg/L), total solids (43,635 mg/L), suspended solids (19,020 mg/L), and oil and grease (8370 mg/L) (Ma, 1995).

Direct release of untreated POME into aquatic environments will cause serious pollution problems. Hence, the treatment requires rapid technology that is efficient to treat POME (Singh et al., 2010). The most conventional method for treatment of POME is using the ponding system (Khalid and Wan Mustafa, 1992; Ma and Ong, 1985). There are other methods such as aerobic and anaerobic digestions, membrane filtration, and physicochemical treatments that might be the answer to the improvement of current POME treatment method.

Sadly, POME treatment that involves biological treatment by anaerobic and aerobic systems is not efficient in treating POME, leading to environmental pollution (Ahmad et al., 2005). Its characteristics of low pH and high BOD values together with the colloidal nature of suspended solids make the treatment by conventional methods difficult (Olie and Tjeng, 1972; Stanton, 1974). The conventional system of POME treatment such as the ponding system is not really effective and is costly. It lowers the profit, restricts the utilization of renewable resources, and also gives rise to the highest environmental pollution proved by calculations done by Indonesia (Schuchardt et al., 2005).

POME has a very high concentration of proteins, carbohydrates, nitrogenous compounds, lipid, and minerals. This leads to the possibilities of reusing it for fermentation media (Habib et al., 1997; Hwang et al. 1978; Phang, 1990; Suwandi, 1991; Wu et al., 2006). According to Wu et al. (2007), POME can be used as fermentation media for producing value-added product such as antibiotics, solvents, bio-insecticides, polyhydroxyalkanoates organic acids, and also enzymes, to various degrees of success. POME is also believed to contain water-soluble antioxidants that are very powerful, phenolic acids, and flavonoids according to Wattanapenpaiboon and Wahlqvist, (2003). Table 2.1 shows the percentage of major constituents, amino acids, fatty acids, and minerals in raw POME (Habib et al., 1997). These characteristics may be the reason for the inhibition of microorganism growth occurring in POME (Lin et al., 2005; Uzel et al., 2005).

2.4 Itaconic Acid

Itaconic acid is known as methylene butanedioic acid or methylene succinic acid. It is an unsaturated dicarboxylic acid produced by the filamentous

TABLE 2.1

Percentage of Major Constituents, Amino Acids, Fatty Acids, and Minerals in Raw POME

Major Constituents	Composition (%)	Amino Acids	Composition (%)	Fatty Acids	Composition (%)	Minerals	Composition (μg/g Dry Weight)
Moisture	6.99	Aspartic acid	9.66	Caprylic acid (8:0)	2.373	Fe	11.08
Crude protein	12.75	Glutamic acid	10.88	Capric acid (10:0)	4.29	Zn	17.58
Crude lipid	10.21	Serine	6.86	Lauric acid (12:0)	3.22	P	14,377.38
Ash	14.88	Glycine	9.43	Myristic acid (14:0)	12.66	Na	94.57
Carbohydrate	29.55	Histidine	1.43	Pentadecanoic acid (15:0)	2.21	Mg	911.95
Nitrogen-free extract	26.39	Arginine	4.25	Palmitic acid (16:0)	22.45	Mn	38.81
Total carotene	0.019	Threonine	2.58	Heptadecanoic acid (17:1)	1.39	K	8,951.55
Total	100.789	Alanine	7.70	10-Heptadecanoic acid (17:1)	1.12	Ca	1,650.09
		Proline	4.57	Stearic acid (18:0)	10.41	Co	2.40
		Tyrosine	3.16	Oleic acid (18:1n–9)	14.54	Cr	4.02
		Phenylalanine	3.20	Linoleic acid (18:2n–6)	9.53	Cu	10.76
		Valine	3.56	Linolenic acid (18:3n–3)	4.72	Ni	1.31
		Methionine	6.88	Y-linolenic acid (18:3n–6)	0	S	13.32
		Cystine	3.37	Arachidic acid (20:0)	3.56	Se	12.32
		Isoleucine	4.53	Eicosatrienoic acid (20:3n–6)	2.04	Si	10.50
		Leucine	4.86	Eicosatetraenoic acid (20:4n–6)	1.12	Sn	2.30
		Lysine	2.66	Eicosapentaenoic acid (20:5n–3)	0.36	Al	16.60
		Tryptophan	1.26	Total	95.99	B	7.60
		Total	90.84			Mo	6.45
						As	9.09
						V	0.12
						Pb	5.15
						Cd	0.44

Source: Adapted from Habib, M. A. B. et al., 1997. *Aquaculture* 158: 95–105.

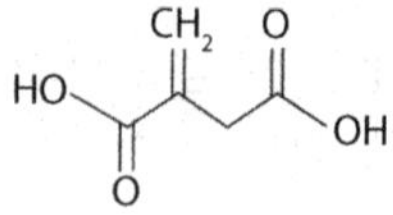

FIGURE 2.1
Itaconic acid structure. (Adapted from Isiklan, N., Kursun, F. 2013. *Polymer Bulletin* 70: 1065–1084.)

fungi *Aspergillus terreus* and *Aspergillus itaconicus* from carbohydrates like sucrose, glucose, and xylose (Kautola, 1990; Reddy and Singh, 2002; Willke and Vorlop, 2001). The synthesis of itaconic acid is uneconomical due to the high cost of substrate that produces relatively low yield (Berg and Hetzel, 1978; Blatt, 1943; Chiusoli, 1962).

The structure of itaconic acid as seen in Figure 2.1 has a methylene group and two carboxylic groups which make it a very useful compound. The double bond is able to participate in polymerization. Itaconic acid can be considered as an R-substituted acrylic acid. It has a high potential as a replacement of acrylic acid or methacrylic acid in polymers. It has various functions as plastics, adhesives, elastomers, and coating when in its polymerized form such as methyl, ethyl, or vinyl esters. Itaconic acid is also used as polyacrylonitrile and styrene–butadiene copolymers (Kroschwitz, 1997). There are also new findings that state that itaconic acid might be useful in the making of artificial glass (Kin et al., 1998), in bioactive compounds in agriculture, pharmaceuticals, and also in medicine (Bagavant et al., 1994). Since it is a versatile starting material, itaconic acid creates new ways to form many useful polyfunctional building blocks by the enzymatic transformations according to Ferraboschi et al. (1994).

The other possible application of itaconic acid derivatives is alkali salts of the homo polymer of itaconic acid which have been recommended for use in detergents (Lancashire, 1969). The application of concentration up to 100 ppm of itaconic acid may be used as a scale inhibitor in boilers according to Walinsky (1984). The imidazole derivative of itaconic acid also has been regarded as an active component in shampoos (Christiansen, 1980). It also competes along with fumaric or malic acid (Smith et al., 1974). Total market size of itaconic acid is considered about 10,000–15,000 tonnes per year. Contributing to the higher part of this market are the polymers, along with another small portion of additives (Gordon and Coupland, 1980), detergents (Lancashire, 1969), and biologically active derivatives involving pharmaceuticals and in agriculture (Bagavant et al., 1994; Karanov et al., 1989).

The itaconic acid synthesis pathway is less understood compared to citric acid. There are three suggested pathways for the synthesis as in Figure 2.2. The first one is from *cis*-aconitate (Bentley and Thiessen, 1957), the second is from citramalate (Jaklitsch et al., 1991; Nowakowska-Waszczuk, 1973) and the third is from 1,2,3-tricar-boxypropane (Shimi and Nour El Dein, 1962).

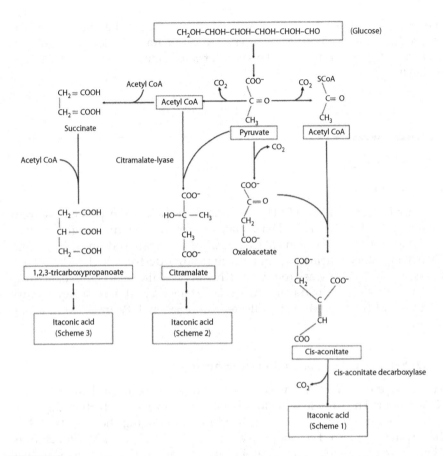

FIGURE 2.2
Pathway of itaconic acid synthesis. (Adapted from Bonnarme et al., 1995. *Journal of Bacteriology* 177: 3573–3578.)

The most possible pathway is the first one. Bentley and Thiessen (1957) proposed that the TCA cycle was active, giving *cis*-aconitate which was then decarboxylated by a crude enzyme preparation from *A. terreus* NRRL 1960 to itaconate and CO_2. The enzyme named *cis*-aconitate decarboxylase (CAD), that gives stoichiometric amounts of itaconate and carbon dioxide, has been partially purified and characterized. Experiments with whole-cell cultures of *A. terreus* done by Bonnarme et al. (1995) provide clear evidence that itaconate synthesis happens via the Krebs cycle. Most of TCA cycle intermediates including cis-aconitic acid were identified in the culture of *A. terreus* and their concentrations are higher in cultures that produced a large amount of itaconic acid (Bonnarme et al., 1995).

Aspergillus terreus NRRL 1960 has been chosen because it is the most published strain of its kind (Lockwood and Reeves, 1945). This strain is equivalent to ATCC 10020. It has been isolated by extensive screening from the organism that is able to produce itaconic acid. It has been reported to produce about 45% (w/w) to 54% (w/w) of itaconic acid per 100 g of glucose utilization in 4–6 days of fermentation according to Nelson et al. (1952).

2.5 Case Study

2.5.1 Materials

Polyvinyl alcohol (PVA) 60,000 MW and boric acid were purchased from Merck Schuchardt OHG, Darmstadt, Germany. Sodium alginate was obtained from Kanto Chemical Co., Inc., Tokyo, Japan, sodium sulfate from GCE Laboratory Chemicals, and calcium chloride from R&M Marketing, Essex, UK. Glycerol was from Qrec Chemical, Malaysia. *Aspergillus terreus* NRRL 1960 was obtained from the United States Department of Agriculture (USDA) and fresh POME was collected from Sedenak Palm Oil Mill, Kulai, Johor.

2.5.2 Sample Collection and Pretreatment

The sample of raw POME was collected from the Sedenak Palm Oil Mill, Kulai, Johor. The sample was stored at 4°C to prevent any further degradation by bacteria. POME was pretreated by centrifuging the sample at 4°C, 4000 rpm for 15 min. The sample was autoclaved at 15 p.s.i, 121°C for 20 min. Then it was centrifuged again at 4°C, 4000 rpm for 15 min. This step was performed to separate the suspended solids in the POME.

2.5.3 Organism and Inoculum Preparation

Aspergillus terreus NRRL 1960 was stored in potato dextrose broth at 4°C until further usage and subculturing the culture on potato dextrose agar is being done every month. The culture was subcultured on potato dextrose agar and incubated at 37°C for 5 days. The spore was then harvested and collected in 10 mL 1% (v/v) Tween 80. The solution was then centrifuged at 4°C, 4000 rpm for 15 min. The supernatant was then discarded and the pellet was resuspended with 10 mL of sterile distilled water and stored at 4°C until further use.

2.5.4 Beads Preparation

Inoculum of *A. terreus* NRRL 1960 10 mL was mixed with 90 mL of 12% (w/v) PVA and 1% (w/v) sodium alginate solution. The mixed solution of

PVA alginate and the inoculum of *A. terreus* were dropped into a mixed solution of 100 mL 5% (w/v) boric acid and 2% (w/v) calcium chloride using a syringe to form beads. The solution was continuously stirred for 50 min on a stirring machine. The PVA–alginate sulfate beads were then kept at 4°C for 24 h. The beads were then strained, washed, and stirred in 10% (w/v) boric acid solution for 30 min and further treated with 0.5 M sodium sulfate solution for another 30 min according to Mohd Zain et al. (2010). The beads were kept at 4°C for further use. The average weight of the beads obtained was 113.43 g and the average bead size was 0.3 cm.

2.5.5 Medium Preparation

The production medium was prepared according to Jamaliah et al. (2006). Pretreated POME (112 mL of 51% (v/v)) was added to 88 mL of 44% (v/v) glycerol and the pH adjusted to 5.8 by adding 1 mol/L hydrochloric acid or 1 mol/L sodium hydroxide. Then 2 g of ammonium nitrate is added to the medium.

2.5.6 Itaconic Acid Fermentation

Fermentation was carried out using a fermentation medium consisting of 51% (v/v) pretreated POME, 44% (v/v) glycerol, and 5% (v/v) distilled water. Four sets of flasks were prepared consisting of 0% (w/v) beads for control, and the experimental part of 5% (w/v) beads, and 10% (w/v) beads containing the *A. terreus* NRRL 1960 and free cells of *A. terreus* NRRL 1960. The experiment was carried out in four replications. The flasks were then incubated in the incubator shaker at 30°C, 150 rpm for 6 days and the sampling process was carried out at the same time starting day 0 until day 6 (Jamaliah et al., 2006).

2.5.7 Determination of pH

The pH of the solution is determined using the pH meter. pH meter Sartorius PB-10 was calibrated before usage. The pH is determined by dipping the pH rod inside the solution.

2.5.8 Determination of Biomass

2.5.8.1 Determination of Biomass in Medium

Biomass determination was done by measuring the dry cell weight. The 0.45 μm nylon filter paper was dried overnight at 70°C and preweighted. Then 10 mL of sample was then filtered through the paper followed by washing twice with distilled water. It was then dried at 70°C until it achieved constant weight (Jamaliah et al., 2006). The biomass (X) is obtained from subtracting weight after by weight before and dividing the result by 10 before multiplying it by 1000.

$$\text{Biomass}\,(X) = \frac{\text{Weight after} - \text{Weight before}}{10} \times 1000 \qquad (2.1)$$

where
  X = biomass (g)
  Weight before = weight of filter membrane only
  Weight after = weight of filter membrane + weight of biomass

2.5.8.2 Determination of Biomass in Immobilized Beads

Bead samples (10 beads) were collected each day of the experiment. The samples were then centrifuged at 5000 rpm for 15 min and washed thrice with distilled water. Then, the 0.45 µm nylon membrane filter containing beads and cell biomass was dried in the oven at 70°C for around 24 h. The filtered membrane was then weighed until constant weight was obtained. The constant weight is the overall mass produced. The mass concentration of the parameters in each sample was measured (Wang and Hu, 2007).

2.5.9 Determination of Itaconic Acid Concentration[*]

Itaconic acid concentration was measured using HPLC using Rezex ROA—Organic Acid column with 300 mm length and 7.8 mm inner diameter. The mobile phase of the column used was 0.005 N sulfuric acid with flow rate of 0.5 mL/min. The temperature was set to be at 75°C. The column was attached to the HPLC Agilent 1100 system which is an auto sampler and equipped with a UV–Vis detector. The itaconic acid was detected at 210 nm wavelengths.

2.5.10 Determination of Glycerol Concentration[*]

HPLC was used for glycerol determination. Rezex ROA—organic acid column with 300 mm length and 7.8 mm inner diameter was used. The mobile phase used was 0.005 N of sulfuric acid with flow rate of 0.6 mL/min, the temperature was set to be at 79.5°C, and the column was attached to the HPLC Agilent 1100 system which uses the auto sampler and is equipped with RID detector.

---

[*] Modified from Phenomenex (2012).

2.5.11 Characterization of PVA–Alginate–Sulfate Beads

2.5.11.1 Determination of Beads Leakage

Phosphate buffer pH 5 was used to store the beads. On the first day, the sample was collected every 6 h. On the second day, the sample was collected every 12 h and every 24 h on each of the subsequent days after (Wong and Azila, 2008). The sample was then analyzed for protein determination using the Lowry Method (Lowry et al., 1951). Firstly, 0.1 mL of 2N NaOH was added to 0.1 mL of sample. It was then hydrolyzed in a water bath at 100°C for 10 min. The sample was then cooled to room temperature and 1 mL of freshly prepared reagent (2% (w/v) $NaCO_3$ in distilled water, 1% (w/v) $CuSO_4 \cdot 5H_2O$ in distilled water and 2% sodium potassium tartrate in distilled water in proportion of 100:1:1) was then added. Lastly 0.1 mL of Folin reagent was added and the mixture was left at room temperature for 30–60 min before the absorbance was read using UVmini 1240 UV–VIS Spectrophotometer (Shimadzu Corporation, Japan) at 750 nm.

2.5.11.2 Determination of Beads Leakage Using FESEM Analysis

To increase the validity of this experiment, the preparation of sample for FESEM analysis was duplicated (Mohd Zain et al., 2010).

The PVA–alginate–sulfate beads were patted dry with tissues and cut with a surgical knife to obtain the cross section of the beads. The beads were then placed on the stand and cross-sectional images were obtained using FESEM (Model Ziess SUPRA 35 VP FESEM). For the dry sample, the beads were cut with a surgical knife and dried in desiccator until a constant reading was obtained. They were then placed on a sample stub and coated four times with platinum coating with Auto-Fine Coater JFC-1600 (Joel, USA Inc., USA). They was then placed on the stand and the images were obtained using FESEM (Model Ziess SUPRA 35 VP FESEM).

2.6 Results and Discussions

2.6.1 Production of Itaconic Acid and Its Kinetics Study

Immobilization does affect the production of itaconic acid. According to the Figure 2.3 below, the highest production of all 5% (w/v) beads, 10% (w/v) beads, and free cells happens on day 3 of fermentation. The highest production of itaconic acid was 9.656 g/L achieved by 10% (w/v) beads, followed by 7.009 g/L achieved by 5% (w/v) , and lastly 3.43 g/L achieved by the free cells of the *A. terreus* NRRL 1960. The amount of spore is set to be the same, that is 2.071×10^8 spore/mL in 10% (w/v) beads and 10% (v/v) inoculum of the free cell fungus so that the results can be easily compared to each other.

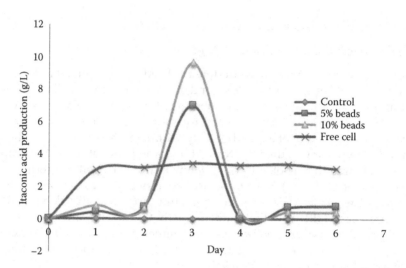

FIGURE 2.3
Itaconic acid production for control, 5% (w/v) beads, 10% (w/v) beads and free cells for 6 days.

The result shows that immobilization affects itaconic acid production by increasing the production up to nearly three times, compared to 10% (w/v) beads and free cells. Production of 10% (w/v) beads and free cells are 9.565 and 3.43 g/L, respectively. The 5% (w/v) and 10% (w/v) beads produced two types of biomass, immobilized biomass and free mycelia biomass found in the medium originated from the leakage cells out of the beads, respectively. Itaconic acid synthesis was suppressed; as the growth of *A. terreus* was inhibited by high concentration of itaconic acid as reported by Kobayashi and Nakamura (1964) and Jaklitsch et al. (1991). Thus, we can assume that the itaconic acid was mainly produced by the immobilized cells that were entrapped in the immobilization matrix and thus not directly interacting with the itaconic acid in the medium. The production of itaconic acid from the immobilized fungus is high compared to free cells resulting from the growth rate in immobilized fungus which is relatively low. Growth of biomass in immobilization is limited depending on the size of space inside the matrix because the resulted biomass will remain immobilized (Kuek, 1986).

Itaconic acid production is one of growth dissociate synthesis. This concluded that if the growth rate is high, the fungus will be focusing on the aggregation of the cells and mycelia and less energy will be used to produce itaconic acid. The purpose of immobilization is to increase itaconic acid production by inhibiting its growth since a growth rate close to zero is preferable for maximal synthesis (Ryu et al., 1979). However, the itaconic acid production decreases rapidly for the immobilized fungus on the day 4 of fermentation. This was not happening in the free cell production based on Figure 2.3.

"The "loss" of the product" may have resulted from the reaction between the alginate (component of the PVA–alginate–sulfate beads) and the itaconic acid itself. This statement is supported by a previous study done by Isiklan and Kursun (2013) where the graft copolymers of sodium alginate and itaconic acid can happen by free radical polymerization using the initiator pair of ammonium nitrate as the redox system. The ammonium nitrate was accidentally as one of the components in the production medium, thus it contributes to the reaction of the component inside the PVA–alginate–sulfate beads and the product of the fermentation, which is itaconic acid.

The reaction between itaconic acid and sodium alginate caused by the *in situ* esterification and grafting reaction studied by Isiklan and Kursun (2013) shows that itaconate contained two polar functional groups (COOH) and could be reacting with OH groups from the sodium alginate chain to form ester bonds in acid catalysis (Isiklan and Kursun, 2013). It is an unsaturated dicarboxylic acid with one carboxyl group conjugated to the methylene group. It is easily copolymerized and provided polymer chains with carboxylic side groups, which are highly hydrophilic and able to form a hydrogen bond with corresponding groups. Small amount of itaconic acid in the polymeric network can increase the degree of swelling (Yu et al., 2011). This might be the reason why the itaconic acid level in the medium suddenly decreased after optimum production on day 3.

The specific growth rate, μ, is obtained from the slope of the biomass production graph for the 5% (w/v) beads, 10% (w/v) beads, and free cells. The specific growth rate for 5% (w/v) beads, 10% (w/v) beads, and free cells are 0.1055, 0.1348, and 0.4588 h^{-1}, respectively. The generation time of the organism can be obtained by dividing in 2 with the specific growth rate, μ. The results for 5% (w/v) beads, 10% (w/v) beads, and free cells are 6.570, 5.1420, and 1.510 h, respectively. This evidence supports that immobilization affects the growth rate by slowing it down. The doubling time of free cells is about four times faster than immobilized cells when compared to the doubling time of both 5% (w/v) beads and 10% (w/v) beads.

The yield coefficient of cells, which is the amount of biomass formation over the glycerol being consumed are different in 5% (w/v) beads, 10% (w/v) beads, and free cells. The yield coefficients are 0.2174, 0.3700, and 0.5875 g/g for 5% (w/v), 10% (w/v), and free cells, respectively. The immobilized *A. terreus* yield coefficients are rather lower than free cells. This might be resulting from immobilization which slows the growth rate in immobilized cells. The substrate being consumed was used to produce the product rather than being used for cell aggregation and growth. That is the reason that a growth rate relatively close to zero is needed for maximal synthesis (Ryu et al., 1979). The yield coefficient of product means the product formation over each gram of glycerol being consumed. The yield coefficient of product of 5% (w/v) beads, 10% (w/v) beads, and free cells are 0.4103, 0.8238, and 0.2632 g/g, respectively. For the free cells, the result is similar to the experiment by Eimhjellen and Larsen (1955), where 23% of itaconic acid was produced from the glycerol consumed. This means that for each gram of glycerol used, itaconic acid produced is 0.23 g, while the 5% (w/v) beads and 10% (w/v) beads produced 0.4103 and 0.8238 g, respectively. This means that for the 5% (w/v) beads it produced near twice the yield and for the 10% (w/v) beads, it produced thrice the yield obtained by the free cells.

The product formation of growth (P/X) from biomass of the 5% (w/v) beads, 10% (w/v) beads, and free cells are 0.7494, 0.8972, and 0.4711 g/g, respectively. This concludes that immobilized fungus can produce higher amount of itaconic acid compared to free cells per gram of biomass. For example, 1 g of biomass of the 10% beads can produce nearly twice more itaconic acid than the free cells' biomass. The productivity for the 5% (w/v) beads, 10% (w/v) beads, and free cells are 0.0486, 0.0670, and 0.023 g/L/h.

2.6.2 Characterization of Outer Layer of PVA–Alginate–Sulfate Bead Using FESEM before and after the Treatment

Figure 2.4 shows outer layer of the PVA–alginate–sulfate beads before treatment with 2500× magnification. No pores are detected on the surface of the beads because the beads must be dry in order to be analyzed under the FESEM.

Figure 2.5 is how the picture of a wet sample should look like. The pores can be seen clearly on the surface of the beads with 150× magnification. The pore is important to maintain the cells viability for mycelia aggregation.

Figure 2.6 shows the mycelium and spore of *A. terreus* aggregates on the surface of PVA–alginate–sulfate beads after treatment under 1000× magnification. This indicates that immobilization of *A. terreus* in PVA–alginate–sulfate

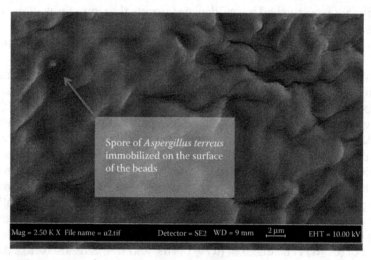

FIGURE 2.4
Outer layer of PVA–alginate–sulfate beads before treatment.

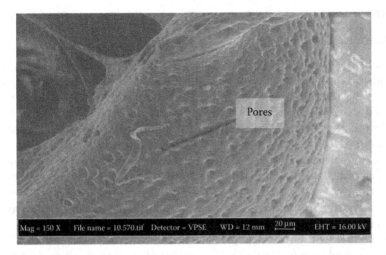

FIGURE 2.5
Example of FESEM picture of outer layer of PVA–alginate sulfate bead with 150× magnification. (Adapted from Mohd Zain, N. A., Suhaimi, M. S., Idris, A. 2010. *Biochemical Engineering Journal* 50: 83–89.)

beads does not affect cell viability. It also proves that the matrix permits the growth of the cell to form the mycelium and thus permits the aggregation of the mycelia of the fungus on the surface of the beads. It also proved that cell leakage occurred in the experiment with the immobilized fungus where the mycelium and the spore can be clearly seen in Figure 2.6.

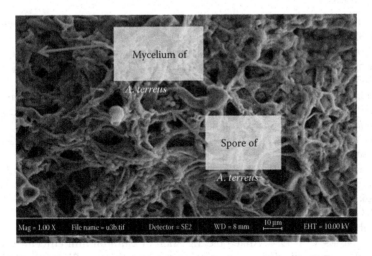

FIGURE 2.6
Mycelia of fungus *A. terreus* NRRL 1960 found on the outer layer of PVA–alginate beads after treatment.

2.7 Conclusion

This study concluded that *A. terreus* NRRL 1960 was successfully immobilized inside the PVA–alginate–sulfate beads, but the immobilization process is not really effective as leakage of the cells occurs and thus the mycelium can be found inside the medium. This is supported by the FESEM picture taken and the biomass determination from the medium that proved the mycelium leaked out of the beads and grew inside the medium.

The characterization of POME was done and proved that the production medium formulation obtained from Jamaliah et al. (2006) is not suitable for POME treatment. This is caused by the addition of 44% (v/v) glycerol to the medium and the immobilized fungus cannot utilize such a large amount of carbon source provided, proven by the high COD value of the medium after day 6 of treatment with 10% (w/v) beads.

Itaconic acid production can be increased by nearly threefold using immobilized *A. terreus* NRRL 1960. But the yield is not really high as glycerol was being used as cosubstrate with POME and the medium formulation of this experiment is not as ideal as conventional substrates such as glucose and sucrose. Glucose, for example, has been reported to produce as much as 80 g/L of itaconic acid produced by the same strain that has been used in this experiment which is the *A. terreus* NRRL 1960.

References

Ahmad, A. L., Ismail, S., Bhatia, S. 2003. Water recycling from palm oil mill effluent (POME) using membrane technology. *Desalination* 157: 87–95.

Ahmad, A. L., Ismail, S., Bhatia, S. 2005a. Ultrafiltration behavior in the treatment of agroindustry effluent: Pilot scale studies. *Chemical Engineering Science* 60: 5385–5394.

Bagavant, G., Gole, S. R., Joshi, W., Soni, S. B. 1994. Studies on anti-inflammatory and analgesic activities of itaconic acid systems. Part 1. Itaconic acids and diesters. *Indian Journal of Pharmaceutical Sciences* 56: 80–85.

Basiron, Y., Weng, C. K. 2004. The oil palm and its sustainability. *Journal Oil Palm Resource* 16(1): 1–10.

Bentley, R., Thiessen, C. P. 1957. Biosynthesis of itaconic acid in *Aspergillus terreus*. I. Tracer studies with ^{14}C-labelled substrates. *Journal of Biology and Chemistry* 226: 673–687.

Berg, R. G., Hetzel, D. S. 1978. US Patent 4100179.

Blatt, A. H. 1943. *Organic Syntheses*, vol 11. Wiley, New York, p. 328.

Borja, R., Banks, C. J. 1994. Anaerobic digestion of palm oil mill effluent using an up-flow anaerobic sludge blanket (UASB) reactor. *Biomass Bioenergy* 6: 381–389.

Bonnarme, P., Gillet, B., Sepulchre, A. M., Role, C., Beloeil, J. C, Ducrocq, , C. 1995. Itaconate biosynthesis in *Aspergillus terreus*. *Journal of Bacteriology* 177: 3573–3578.

Chiusoli, 1962. G. P. US Patent 3025320.

Choo Yuen May. 2014. MPOB Menjana Perubahan Industri Sawit. *Berita Harian* 2.

Christiansen, A. 1980. GB-Patent 1 574 916 (to Miranol Chemical): Surface Active Amide and Amideazolines. 31: 509–521.

Dalimin, M. N. 1995. Renewable energy update: Malaysia. *Renewable Energy* 6(4): 435–439.

Davis, J. B., Reilly, P. J. A. 1980. Palm oil mill effluent—A summary of treatment methods. *Oleagineux* 35: 323–30.

Eimhjellen, K. E., Larsen, H. 1955. The mechanism of itaconic acid formation by *Aspergillus terreus* 1. The effect of acidity. *Biochemistry Journal* 60(1): 135–139.

Ferraboschi, P., Casati, S., Grisenti, P., Santaniello, E. 1994. Selective enzymatic transformations of itaconic acid derivatives: An access to potentially useful building blocks. *Tetrahedron* 50: 3251–3258.

Foo, K. Y., Hameed, B. H. 2010. Insight into the applications of palm oil mill effluent: A renewable utilization of the industrial agricultural waste. *Renewable Sustainable Energy Review* 14: 1445–1452.

Gordon, A. A., Coupland, K. 1980. DE-Patent 3 001 000 (to Exxon Research and Engineering): Mehrzweckschmiermittel.

Habib, M. A. B., Yusoff, F. M., Phang, S. M., Ang, K. J., Mohamed, S. 1997. Nutritional values of chironomid larvae grown in palm oil mill effluent and algal culture. *Aquaculture* 158: 95–105.

Hwang, T. K., Ong, S. M., Seow, C. C., Tan, H. K. 1978. Chemical composition of palm oil mill effluents. *The Planter* 54: 749–756.

Isiklan, N., Kursun, F. 2013. Synthesis and characterization of graft copolymer of sodium alginate and poly(itaconic acid) by the redox system. *Polymer Bulletin* 70: 1065–1084.

Jaklitsch, W. M., Kubicek, C. P., Scrutton, M. C. 1991. The subcellular organization of itaconate biosynthesis in *Aspergillus terreus*. *Journal of General Microbiology* 137: 533–539.

Jamaliah, M. J., Noor, I. S. M., Wu, T. Y. 2006. Factor analysis in itaconic acid fermentation using filtered POME by *Aspergillus terreus* IMI 282743. *Jurnal Kejuruteraan* 18: 39–48.

Kalam, M. A., Masjuki, H. H. 2002. Biodiesel from palm oil—An analysis of properties and potential. *Biomass Bioenergy* 23: 471–479.

Karanov, E. N., Georgiev, G. T., Mavrodiev, S. I., Aleksieva, V. S. 1989. Derivatives of some aliphatic dicarboxylic acids: Their influence on vegetative growth of fruit trees. *Acta Horticurae* 239: 243–248.

Kathiravale, S., Ripin, A. 1997. Palm oil mill effluent treatment towards zero discharge. A Paper Presented at National Science and Technology Conference, Kuala Lumpur, Malaysia (July, 15th–16th 1997), pp. 1–8.

Kautola, H. 1990. Itaconic acid production from xylose in repeated-batch and continuous bioreactors. *Applied Microbiology Biotechnology* 33: 7–11.

Khalid, A. R., Wan Mustafa, W. A. 1992. External benefits of environmental regulation resource recovery and the utilization of effluents. *Environmentalist* 12: 277–285.

Kin, R., Sai, T., So, S. 1998. Itaconate copolymer with quadratic nonlinear optical characteristic. JP Patent 10,293,331.

Kobayashi, T., Nakamura, I. 1964. Dynamics in mycelia concentration of *A. terreus* K26 in steady state of continuous culture. *Journal of Fermentation Technology* 44: 264–274.

Kroschwitz, J. 1997. Itaconic acid. In: Home-Grant, M. (Ed.), *Kirk-Othmer Encyclopedia of Chemical Technology*. 4th ed., John Wiley and Sons: New York, Vol. 14, p. 952.

Kuek, C. 1986. Immobilized living fungal mycelia for the growth dissociated synthesis of chemicals. *International Industrial Biotechnology*. 56(6): 4–5.

Lancashire, E. 1969. US-Patent 3 454 500 (to Procter and Gamble): Soap compositions having improved curd-dispersing properties.

Lin, Y. T., Labbe, R. G., Shetty, K. 2005. Inhibition of *Vibrio parahaemolyticus* in seafood systems using oregano and cranberry phytochemical synergies and lactic acid. *Innovative Food Science Emerging Technology* 6: 453–458.

Lockwood, L. B., Reeves, M. D. 1945. Some factors affecting the production of itaconic acid by *Aspergillus terreus*. *Archieve of Biochemistry Biophysics* 6: 455–469.

Lowry, O. H., Rosebrough, N. J., Farr, A. L., Randall, R. J. 1951. Protein measurement with the folin phenol reagent. *Journal of Biological Chemistry* 193: 265–275.

Ma, A. N. 1995. A novel treatment for palm oil mill effluent. *Palm Oil Research Institute Malaysia* (PORIM); 29: 201–12.

Ma, A. N. 2000. Environmental management for the palm oil industry. *Palm Oil Development* 30: 1–9.

Ma, A. N., Cheah, S. C., Chow, M. C. 1993. Current status of palm oil processing wastes management. In: Yeoh, B. G., Chee, K. S., Phang, S. M., Isa, Z., Idris, A., Mohamed, M. (Eds.), *Waste Management in Malaysia: Current Status and Prospects for Bioremediation*. Ministry of Science, Technology and the Environment: Malaysia, pp. 111–136.

Ma, A. N., Ong, A. S. H. 1985. Pollution control in palm oil mills in Malaysia. *Journal of American Oil Chemists' Society* 62: 261–266.

Mohd Zain, N. A., Suhaimi, M. S., Idris, A. 2010. Hydrolysis of liquid pineapple waste by invertase immobilized in PVA–alginate matrix. *Biochemical Engineering Journal* 50: 83–89.

MPOB. 2008. Number of mills and capacity: 2008. http://econ.mpob.gov.my/economy/annual/stat2008/ei_processing08.htm (accessed September 2009).

Nelson, G. E. N., Traufler, D. H., Kelley, S. E., Lockwood, L. B. 1952. Production of itaconic acid by *Aspergillus terreus* in 20-liter fermentors. *Industrial Engineering Chemistry* 44: 1166–1168.

Nowakowska-Waszczuk, A. 1973. Utilization of some tricarboxylic-acid-cycle intermediates by mitochondria and growing mycelium of *Aspergillus terreus*. *Journal of General Microbiology* 79: 19–29.

Olie, J. J., Tjeng, T. D. 1972. Treatment and disposal of wastewater from a palm oil mill. *Oleagineux* 27: 215–218.

Phang, S. M. 1990. Algal production from agro-industrial and agricultural waste in Malaysia. *Ambio* 19: 415–418.

Phenomenex. 2012. Rezex carbohydrate, oligosaccharides and organic acid separations. https://phenomenex.blob.core.windows.net/documents/6f28134c-e141-44d3-b884-9f0ac38b47eb.pdf (accessed March 2012).

Prasertsan, S., Prasertsan, P. 1996. Biomass residues from palm oil mills in Thailand: An overview on quality and potential usage. *Biomass Bioenergy* 11: 387–395.

Reddy, C. S. K., Singh, R. P. 2002. Enhanced production of itaconic acid from corn starch and market refuse fruits by genetically manipulated *Aspergillus terreus* SKR10. *Bioresource Technology* 85: 69–71.

Reijnders, L., Huijbregts, M. A. J. 2008. Palm oil and the emission of carbon-based greenhouse gases. *Journal of Cleaner Production* 16: 477–482.

Ryu, D., Andreotti, E., Mandels, M., Gallo, B. 1979. Studies on quantitative physiology of *Trichoderma reesei* with two-stage continuous culture for cellulase production. *Biotechnology Bioengineering* 21: 1887–1903.

Schuchardt, F., Wulfert, K., Damoko, D. 2005. New process for combined treatment of waste (EFB) and waste water (POME) from palm oil mills-technical, economical and ecological aspects. *Landbauforsch Volkenrode* 55: 47–60.

Shimi, I. R., Nour El Dein, M. S. 1962. Biosynthesis of itaconic acid by *Aspergillus terreus*. *Archives of Microbiology* 44:181–188.

Singh, R. P., Ibrahim, M. H., Esa, N., Iliyana, M. S. 2010. Composting of waste from palm oil mill: A sustainable waste management practice. *Review in Environmental Science Biotechnology* 9: 331–344.

Smith, J. E., Nowakowska-Waszczuk, A., Anderson, J. G. 1974. Organic acid production by mycelial fungi. In: Spencer, B. (*Ed.*), *Industrial Aspects of Biochemistry*. Elsevier: Amsterdam, pp. 297–317.

Stanton, W. R. 1974. Treatment of effluent from palm oil factories. *The Planter* 50: 382–387.

Suwandi, M. S. 1991. POME, from waste to antibiotic and bioinsecticide. *Jurutera Kimia Malays* 1: 79–99.

Tengku, E.M., Sultan, A.I., Hakimi, M. I. 2012. Vermifiltration of palm oil mill effluent (POME). In: *UMT 11th International Annual Symposium on Sustainability Science and Management*, Terengganu, Malaysia, pp. 1292–1297.

Uzel, A., Sorkun, K., Önçağ, Ö., Çoğulu, D., Gençay, Ö., Sali, H. B. 2005. Chemical compositions and antimicrobial activities of four different Anatolian propolis samples. *Microbiology Research* 160: 189–195.

Vairappan, S. C., Yen, A. M. 2008. Palm oil mill effluent (POME) cultured marine microalgae as supplementary diet for rotifer culture. *Journal of Applied Phycology* 20: 603–608.

Vikineswary, S., Kuthubutheen, A. J., Ravoof, A. A. 1997. Growth of *Trichoderma harzianum* and *Myceliophthora thermophila* in palm oil sludge. *World Journal of Microbiology and Biotechnology* 13: 189–194.

Walinsky, S. W. 1984. US-Patent 4 485 223 (to Pfizer): (Meth) acrylic acid/itaconic acid copolymers their preparation and use as antiscalants.

Wang, B., Hu, Y. 2007. Comparison of four supports for adsorption of reactive dyes by immobilized *Aspergillus fumigatus* beads. *Journal of Environmental Sciences* 19: 451–457.

Wattanapenpaiboon, N., Wahlqvist, M. L. 2003. Phytonutrient deficiency: The place of palm fruit. *Asia Pacific Journal of Clinical Nutrition* 12: 363–368.

Willke, T., Vorlop, K. 2001. Biotechnological production of itaconic acid. *Applied Microbiology Biotechnology* 56: 289–295.

Wong, F. L., Azila, A. A. 2008. Comparative study of poly(vinyl alcohol)-based support materials for the immobilization of glucose oxidase. *Journal of Chemical Technology and Biotechnology* 83: 41–46.

Wu, T. Y., Mohammad, A. W., Md Jahim, J., Anuar, N. 2006. Treatment of palm oil mill effluent (POME) using ultrafiltration membrane and sustainable reuse of recovered products as fermentation substrate. *4th Seminar on Water Management (JSPSVCC)*, Johor, Malaysia, pp. 128–135.

Wu, T. Y., Mohammad, A. W., Jahim, J. M., Anuar, N. 2007. Palm oil mill effluent (POME) treatment and bioresources recovery using ultrafiltration membrane: Effect of pressure on membrane fouling. *Biochemistry Engineering Journal* 35: 309–317.

Yu, Y., Li, Y., Liu, L., Zhu, C., Xu, Y. 2011. Synthesis and characterization of pH and thermoresponsive poly(N-isopropylacrylamideco-itaconic acid) hydrogels crosslinked with N-maleyl chitosan. *Journal of Polymer Research* 18(2): 283–291.

Yusoff, S. 2006. Renewable energy from palm oil innovation on effective utilization of waste. *Journal of Cleaner Production* 14: 87–93.

3

Optimization of Lipid Content in Microalgae Biomass Using Diluted Palm Oil Mill Effluent by Varying Nutrient Ration

Noor Amirah Abdul Aziz, Shreeshivadasan Chelliapan, Mohanadoss Ponraj, Mohd Badruddin Mohd Yusof, and Mohd. Fadhil Md Din

CONTENTS

3.1 Introduction

The oil palm is one of the world's most rapidly expanding equatorial crops. In Malaysia, oil palm plantation currently occupies the largest acreage of farmed land and the palm oil industry is growing rapidly. Malaysia is one of the major palm oil producers in the world (Lam et al., 2009). While the palm oil industry has been recognized strongly for its contribution toward economic growth and rapid development, it has also contributed to environmental pollution due to the production of large quantities of by-products during the process of oil extraction (Parthasarathy et al., 2016). Microalgal culture has received more attention, given its prospects as a source of bioenergy and its potential for wastewater treatment. In this respect, simple and easily cultivated biomass has a number of applications, ranging from its direct use such as biodiesel and various pigments (Fulton, 2004). The complication of cultivation methods and the high cost of growth medium have become a major drawback for the algal industry; nevertheless, the integration along with wastewater treatment has provided a feasible solution due to the fact that exploitation of wastewater as the source of growth medium simultaneously eliminates the requirement for an expensive medium and at also remediates the wastewater.

The current study investigates the potential, benefits, strategies, and challenges of microalgae to be integrated with wastewater treatment, particularly palm oil mill effluent (POME) treatment in Malaysia, which due to the hazardous properties of POME may lead to severe environmental pollution. The integration of POME treatment using microalgal culture will potentially reduce the wastewater treatment retention time and eliminate toxic elements, which serve as nutrients for the growth of microalgae. Moreover, microalgae are gaining considerable attention as a feedstock for lipid production as they can be grown away from the croplands and hence do not compromise food crop supplies (Liu et al., 2008). The optimization of lipid content in microalgal biomass using diluted POME by changing the nutrient ratio is discussed in this study.

3.1.1 Background of Study

The Malaysian palm oil industry is growing rapidly and becoming a very important agriculture-based industry, where the country today is the world's leading producer and exporter of palm oil, replacing Nigeria as the chief producer since 1971. As the world's largest palm oil producer, Malaysia produced 10.6 million tonnes of palm oil in 1999 and increased to 17.7 million tonnes of palm oil in 2008 (MPOB, 2010). This figure is expected to rise as the demand for palm oil increases since it is one of the most important vegetable oils in the world's oil and fats market. The total oil palm planted area in the country in the year 2009 alone was 4.69 million hectares.

Unfortunately, this vital agricultural and industry activity generates a significant amount of by-product known as POME. For every ton of product produced in the extraction process, about 2.5–3.5 tonnes of POME are generated. The effluent is nontoxic as no chemicals are added into the extraction process, and it is fairly acidic with pH ranging from 4.0 to 5.0 as it contains organic acids in complex forms that are suitable to be used as carbon sources (Wu et al., 2010). If the discharged effluent is not properly treated, it can surely cause substantial environmental problems. The palm oil mill industry in Malaysia is identified as the one generating the largest pollution load in rivers throughout the country.

Algae are commonly found in the water system. Algae can be used for wastewater treatment to eliminate organic carbon from wastewater systems. Algae can be categorized into two main groups; microalgae and macroalgae. Microalgae can grow under two conditions: autotrophic and heterotrophic. Autotrophic microalgae are the producers in a food chain, such as plants on land or algae in water. Heterotrophic microalgae are organisms that use organic carbon for growth. A microalga is a photosynthetic microorganism that is able to use solar energy to combine water with carbon dioxide to create biomass (Widjaja et al., 2009). Microalgae have been suggested as a very suitable candidate for fuel production because of their advantages of higher photosynthetic efficiency, higher biomass production, and faster growth compared to other energy crops. Many microalgae strains have been identified capable of producing high content of lipid and most of them are marine microalgae. POME is a carbon source for microalgae that will help the growth of microalgae. Different organic loading will cause different growth of microalgae.

3.1.2 Problem Statement

The palm oil industry is among the main production industries in Malaysia. The palm oil industry is identified as the one generating the largest pollution load in rivers throughout the country. With a thick brownish colloidal mixture of water, oil, and suspended solids, it possesses a very high BOD_3 which is a hundred times more polluting than domestic waste and this adversely affects the water environment. A suitable treatment needs to be worked out since POME has higher potential toward the production of biofuel. This effluent can be used as one of the biofuel resources using microalgae to produce biomass in producing lipid to produce biodiesel. The abundance of POME in Malaysia is an advantage in producing this biofuel.

3.1.3 Scope of Study

This study was conducted to observe the ability of *Chlorella pyrenoidosa* and *Chlorella vulgaris* in diluted POME with concentration of 250 mg/L in different concentrations of nutrient, in terms of lipid production. The microalgae were cultured in Bold's Basal Medium (BBM) for 2 weeks for the acclimatization process before conducting tests. The focus of this study is microalgae produced higher lipid content in POME culture.

3.1.4 Significance of Study

There are a lot of components in the POME substance, that could be useful especially in producing biofuel. Biodiesel is a future product that would be important due to its ability to be used as an alternative fuel. Besides, biodiesel has biodegradable behavior and it is a renewable energy.

3.2 Literature Review

3.2.1 Microalgae

Microalgae known as microphytes are a microscopic algae, which can be found in freshwater and marine systems. They are unicellular cells, which can exist individually, in groups or chains. Mata et al. (2010) have stated that microalgae are prokaryotic or eukaryotic photosynthetic microorganisms that can grow rapidly and live in harsh conditions due to their unicellular or simple multicellular structure.

Microalgae do not look like higher plants. They do not have roots, stems, or leaves but are capable of performing photosynthesis. They use carbon dioxide gas as the food source to grow photoautotrophically.

Mata et al. (2010) stated that microalgae can provide feedstock for several different types of renewable fuels such as biodiesel, methane, hydrogen, ethanol, and bioelectricity. Algae biodiesel contains no sulfur and performs as well as petroleum diesel, while reducing emissions of particulate matter, CO, hydrocarbons, and SO_x. NO_x may be higher in some engine types.

Microalgae have been suggested as a very suitable candidate for fuel production because of their advantages of higher photosynthetic efficiency, higher biomass production, and faster growth compared to other energy crops (Widjaja et al., 2009).

3.2.1.1 Cell Nutrients

For cell protoplasm, algae need a source of carbon other than light. Carbon dioxide is the primary carbon source for algae. Algae grow much better in waters containing high concentrations of bicarbonate alkalinity than in waters with low bicarbonate alkalinity. For algae protoplasm, nitrogen is necessary to form proteins. Ammoniacal nitrogen is the primary source of nitrogen for algae with nitrates as the secondary source. Nitrates must be reduced to ammonia for incorporation into protoplasm. Phosphorus is a critical element for the growth of algae. Although algae do not need a large quantity of phosphorus, it is important in energy transfer for the algae. Phosphates are the primary source of phosphorus for the algae. Since phosphates are limited in the natural environment, phosphorus availability is often the limiting factor in the growth of algae. Algae also need sulfates, trace metals, iron, and magnesium.

3.2.1.2 Growth

The growth of algae follows the same general pattern as the growth of bacteria and fungi. In an excess of light and nutrients, growth is restricted only to the ability of the algae to process the nutrients. Unrestricted growth results in a log increase of cell mass. Algae undergo endogenous respiration. In the presence of light, the nutrients released by endogenous respiration are immediately metabolized back to normal protoplasm. The rate of endogenous respiration is directly proportional to the active mass of algae. Proteins are the primary materials undergoing endogenous respiration and leaving cell wall polysaccharides.

3.2.2 Heterotrophic and Autotrophic Conditions

An organism is heterotrophic when it uses organic carbon for growth. On the other hand, an autotroph organism uses a source of energy such as light to produce organic substrate from inorganic carbon dioxide.

3.2.3 Biodiesel

Biodiesel consists of fatty acid methyl esters originating from vegetable oils and animal fats (Widjaja et al., 2009; Feng et al., 2011). It is a biodegradable, renewable, and nontoxic fuel. Furthermore, it contributes no net carbon dioxide or sulfur to the atmosphere and emits less gaseous pollutants than normal diesel (Widjaja et al., 2009). Biodiesel industries are expanding rapidly both in the United States and Europe with soybean or rapeseed oils as the feedstock (Feng et al., 2011).

Biodiesel is a mixture of fatty acid alkyl esters obtained by transesterification of vegetable oils or animal fats. For biodiesel production, lipids and fatty acids have to be extracted from the microalgal biomass. The most common ways are transesterification as the biodiesel from transesterification can be used directly or as blends with diesel fuel in diesel engines. Biodiesel primarily rapeseed methyl ester, has been in commercial use as an alternative fuel since 1988 in many European countries (Lang et al., 2001).

In particular, biodiesel has two main advantages. First, biodiesel helps in terms of mitigation of the excessive emission of carbon dioxide. Second, biodiesel can be a substitute for petroleum. The key processes involved in biodiesel production using microalgae are cultivation, harvest, lipid extraction (cell disruption), and the transesterification of the lipid (Lee et al., 2010).

3.2.4 Palm Oil Mill Effluent

POME is a waste produced from palm oil processing plants. This oily waste is produced in large volume, and it needs an efficient treatment to avoid environmental hazards. The palm oil industry in Malaysia is growing rapidly and is among the major production industries in Malaysia. Malaysia is one of the world's leading producers and exporters of palm oil. Large quantities of water are used during the extraction process of crude palm oil from the fresh fruit. About 50% of the water results in POME. It is estimated that for 1 tonne of crude palm oil produced, 5–7.5 tonnes of water will end up as POME. In the year 2004, more than 40 million tonnes of POME were generated from more than 370 mills in Malaysia, and this amount keeps increasing. In the year 2008 alone, at least 44 million tonnes of POME were generated from 410 operated mills (Wu et al., 2010). Thian et al. (2010) found that microalgae can be found in POME. Typically with very high organic content and oil, the resulting fresh POME is a thick brownish color liquid. In addition, this type of POME is discharged at a high temperature, between 80°C and 90°C. It possesses a high BOD_3 which is about a hundred times more polluting than domestic waste.

The characteristics of POME may vary considerably for different batches, days, and factories, depending on the processing techniques and the age or type of fruit as well as the discharge limit of the factory, climate, and condition of the palm oil processing (Wu et al., 2010).

3.2.4.1 Treatment of POME

Ponding is the most commonly used system in Malaysia to treat POME. A typical ponding system is as follows: raw POME generated from the milling processes is discharged into the raw pond. Adequate hydraulic retention time (HRT) is provided for the settling process to take place in the raw pond. The raw POME is flowed into an anaerobic pond in which anaerobic degradation occurs, followed by an aeration pond, and then discharged to a watercourse (Neoh et al., 2015).

3.2.4.1.1 Aerobic Digestion

A system using an aerobic digestion for POME treatment would be more efficient and HRT is even shorter than an anaerobic system. Many researchers have found that using an aerobic system, the COD removal in the waste is high; more than 95% can be achieved in a much shorter HRTs than in usual anaerobic digestion.

3.2.4.1.2 Anaerobic Digestion

Anaerobic digestion is the most suitable method for the treatment of effluents containing a high concentration of organic carbon such as POME. The suggested anaerobic treatment process for POME includes the anaerobic suspended growth process, attached growth anaerobic process, anaerobic sludge blanket process, membrane separation anaerobic treatment process, and hybrid anaerobic treatment process. Today, 85% of POME treatment is based on an anaerobic and facultative ponding system, which is followed by another system consisting of an open tank digester coupled with extended aeration in a pond (Vijayaraghavan et al., 2007).

3.2.4.1.3 Bioreactor System

This is a simple and innovative bioreactor process that is capable of treating POME efficiently. The system is superior to the conventional system as it operates with a very short HRT, takes high organic loading, requires less space, and is more environmentally friendly.

3.3 Methodology

Figure 3.1 will shows the methodology of research.

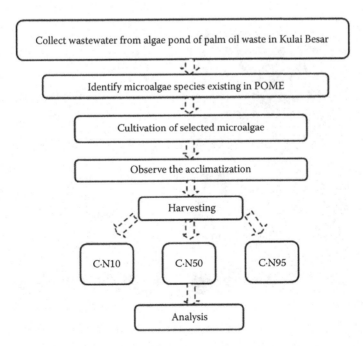

FIGURE 3.1
Research framework.

3.4 Results and Analysis

3.4.1 Identification of Microalgae

A light microscope aided in identifying microalgae species in the culture samples. Slides with microalgae were then observed under the microscope with 10 times magnification. Microalgae existing in POME wastewater were identified. Wastewater for the experiment was collected from an algae pond of palm oil waste in Felda Kulai Besar, Johor.

The observed algae found will then be referred to the nearest morphology described by Bersanti and Gualtieri (2006) in algae dichotomous key identification. The sample most probably consists mainly of *Chlorella pyrenoidosa* as well as *Spirogyra* sp. According to Bersanti and Gualtieri (2006), most of the microalgae sizes observed were to be around 1 μm to a few millimeters only.

Chlorella pyrenoidosa was found to be dominant over *Spirogyra* sp. Adaptation to nutrients from POME has decided the types of dominant microalgae species in a community. The high concentration of organic nutrient available in POME is more favorable for *Chlorella pyrenoidosa* hence more dominant over other species present in the sample. The microalgae found in POME are shown in Figure 3.2a–d.

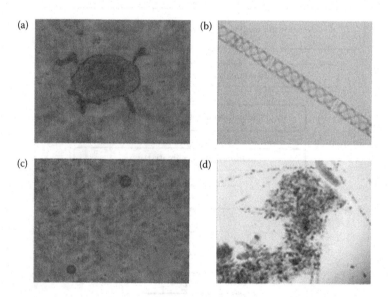

FIGURE 3.2
The microalgae found in POME. (a) *Chlorella*. (b) *Spirulina*. (c) Microalgae cells 10× dilution. (d) Microalgae cells 10× dilution.

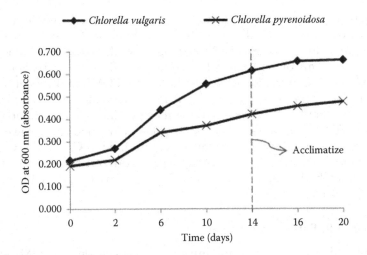

FIGURE 3.3
Chlorella vulgaris and *Chlorella pyrenoidosa* acclimatization.

3.4.2 Acclimatization of Microalgae

Since *Chlorella vulgaris* comes in a synthetic medium, it is of interest to know whether the chlorella can be adapted to POME conditions or not. From the findings, it shows that *Chlorella vulgaris* and *Chlorella pyrenoidosa* can adapt to POME wastewater. As we know, *Chlorella pyrenoidosa* exists in POME, but from the findings, it shows that *Chlorella vulgaris* can grow better than *Chlorella pyrenoidosa*. Figure 3.3 shows at day 10, both the microalgae start to stabilize, and the acclimatization is almost constant.

3.4.3 Determination of Carbon/Nitrogen Ratio (C/N)

A total nitrogen test and COD test was conducted to investigate the carbon/nitrogen ratio. The findings are as shown in Table 3.1. C/N ratio was

TABLE 3.1

Carbon/Nitrogen Composition in Various Ratios

Raw Pome: Algae Pond	COD Total	Total Nitrogen	C/N Ratio
10:90	2275	313	100:14
50:50	32,800	3010	100:9
70:30	27,200	2300	100:8.5
80:20	39,000	4340	100:11
90:10	44,900	3540	100:8
95: 5	57,400	3530	100:6

determined in 250 mg/L concentration of POME in 1 L medium. Using the equation $m_1v_1 = m_2v_2$ the C/N ratio was obtained. The experiment was carried out in 10 mL of raw POME and algae pond mixing. 10:90 ratio indicates that, in 10 mL of the mixing, 1 mL is raw POME and 9 mL is algae pond. Table 3.1 shows that the existence of a lot of raw POME in the solution will obtain lower C/N ratio value while having a limited amount of raw POME, a high concentration of nutrient ratio will be obtained. To determine which variable will be used in the next experiment, higher, middle, and lower C/N ratios are chosen to investigate which concentration of C/N ratio will give higher lipid content. Feng et al. (2011) state that depletion in nutrient ratio will produce higher lipid content and higher nutrient ratio will produce lower lipid content.

3.4.4 Growth Rate of Microalgae

The optical density (OD) test, and mixed liquor suspended solid (MLSS) test were performed to determine the growth rate of *Chlorella pyrenoidosa* and *Chlorella vulgaris*. The variation of the substrate ratio in OD with time for 20 days of batch operation is depicted in Figure 3.4a and b. Both the strains of microalgae achieve optimum growth at 95:05 ratio. The microalgae could not appropriately grow at a 10:90 ratio. It shows a similar trend with MLSS behavior (Figure 3.5a and b).This means that both of the microalgae would be suitable at 95:05 ratio for growth.

3.4.5 Cell Biomass

Cell biomass determination was conducted through chemical and physical tests. For chemical observation, the COD biomass test was conducted, while for the physical test the cell dry weight test was done. Both strains show similar trends in COD biomass and cell dry weight but *Chlorella pyrenoidosa* at ratio 10:90, shows different results. This is due to the human error done

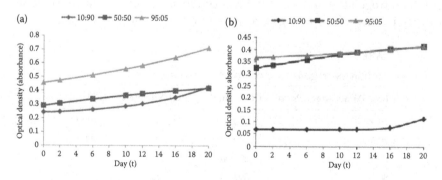

FIGURE 3.4
(a and b) OD of *Chlorella vulgaris* and *Chlorella pyrenoidosa* versus time.

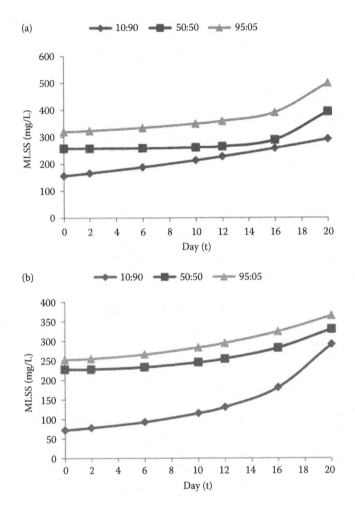

FIGURE 3.5
(a and b) MLSS of *Chlorella vulgaris* and *Chlorella pyrenoidosa* versus time.

when conducting the experiment. Ratio 95:05 for both strains shows a higher starting point and continuously leading the growth, compared to the ratio 10:90. It shows that the ratio is suitable for the strains in producing high biomass thus producing high lipid content (Figures 3.6a and b and 3.7a and b).

3.4.6 Lipid

The gravimetric method has been used to obtain lipid mass production. From the findings obtained, comparison between cell dry weight and lipid has been made. Figure 3.8a shows cell dry weight versus lipid of *Chlorella*

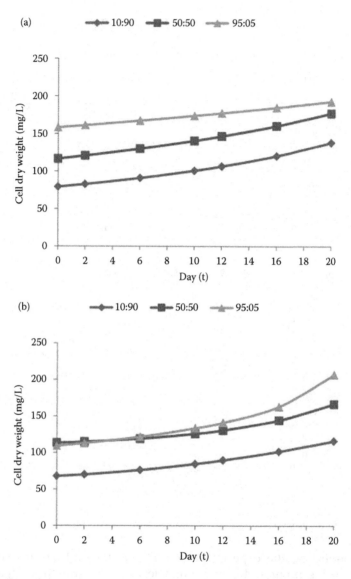

FIGURE 3.6
(a and b) Cell dry weight of *Chlorella vulgaris* and *Chlorella pyrenoidosa* versus time.

vulgaris. It shows that at a ratio of 95:05, the lipid production is higher than the other ratio. At day 20, the cell dry weight obtained was 193 mg/L and the lipid was 56 mg/L compared to ratio 10:90, the cell dry weight obtained is 139 mg/L, a lipid is 40 mg/L and ratio 5:50 shows cell dry weight 177 and 51 mg/L for lipid production.

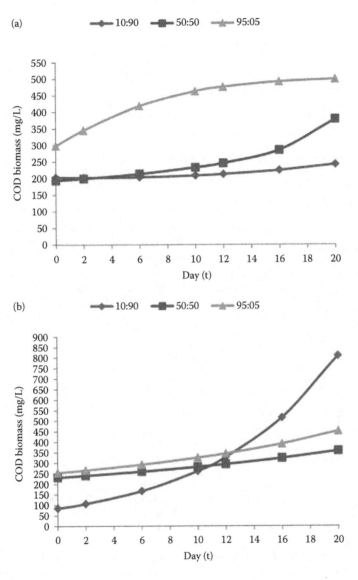

FIGURE 3.7
(a and b) COD biomass of *Chlorella vulgaris* and *Chlorella pyrenoidosa* versus time.

Figure 3.8b indicates cell dry weight versus lipid of *Chlorella pyrenoidosa*. It shows at ratio 95:05, the lipid production is higher than the other ratio. At day 20, the cell dry weight obtained was 207 mg/L, and the lipid was 60 mg/L compared to ratio 10:90. The cell dry weight obtained is 116 and 34 mg/L for lipid production. To sum up, it was concluded that *Chlorella pyrenoidosa* produces higher lipid content compared to *Chlorella vulgaris* at 95:05 ratio.

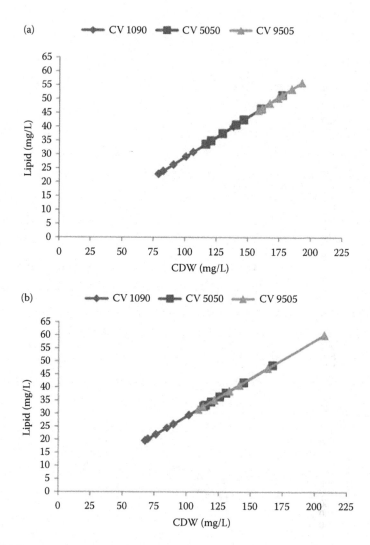

FIGURE 3.8
(a and b) Cell dry weight of *Chlorella vulgaris* and *Chlorella pyrenoidosa* versus lipid.

3.4.7 Ratio MLVSS/MLSS

Ratio of MLVSS/MLSS is another way to know the production of lipid other than quantifying the lipid. It can be used as an early prediction of lipid production for each ratio and strains before the lipid test was done. The analysis above shows that for both strains, ratio 10:90 will produce higher lipid content rather than the 95:05 and 50:50 ratios. It shows that both strains have higher MLVSS value over MLSS. The result can vary when the quantifying lipid test was done. (Figure 3.9a and b, MLVSS/MLSS of *Chlorella vulgaris* and *Chlorella pyrenoidosa* versus time).

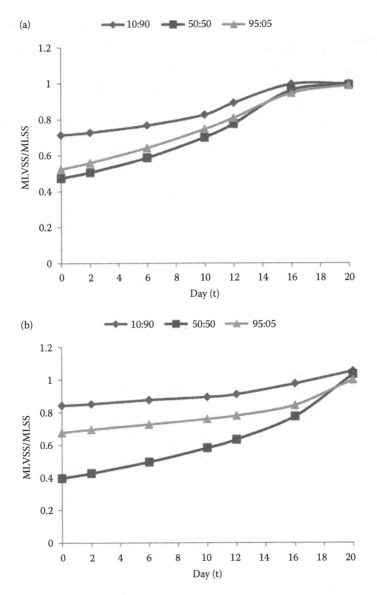

FIGURE 3.9
(a and b) MLVSS/MLSS of *Chlorella vulgaris* and *Chlorella pyrenoidosa* versus time.

3.4.8 Chlorophyll-a, Lipid, and Biomass

Biomass, lipid, and chlorophyll-a are used in analyzing the best strain with best nutrient ratio to determine optimization of lipid content. High biomass results in high production of lipid. Figure lipid shows ratio 95:05 produces high lipid over time. The analysis shows that the lipid production is directly proportional to the biomass generated. Chlorophyll relates to the process of photosynthesis for plants. Ratio 50:50 shows higher absorbance of chlorophyll-a for both strains. Ratio 95:05 for both strains shows lower absorbance of chlorophyll-a. It shows that for ratio 95:05 the chlorophyll used is not for survival but when there is depletion in nutrients, the chlorophyll acts as nutrient for lipid production.

Insert (Figure 3.10a and b Chlorophyll-a of *Chlorella vulgaris* and *Chlorella pyrenoidosa* versus time)

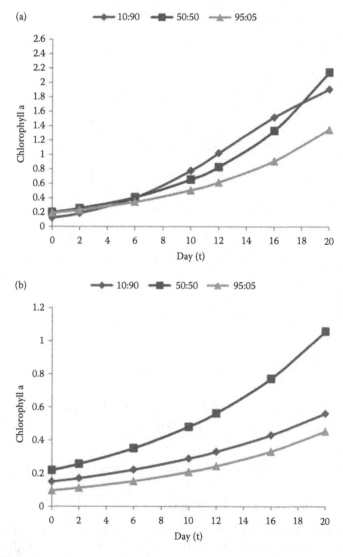

FIGURE 3.10
(a and b) Chlorophyll-a of *Chlorella vulgaris* and *Chlorella pyrenoidosa* versus time.

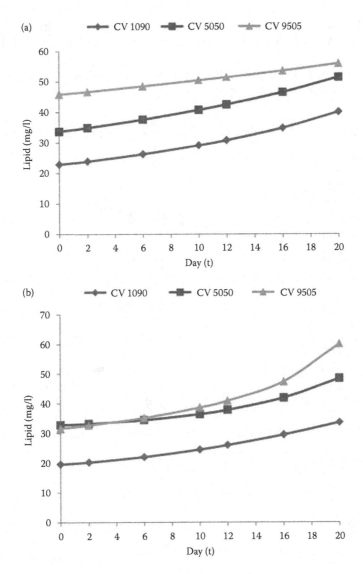

FIGURE 3.11
(a and b) Lipid of *Chlorella vulgaris* and *Chlorella pyrenoidosa* versus time.

Insert (Figure 3.11a and b lipid of *Chlorella vulgaris* and *Chlorella pyrenoidosa* versus time)

3.4.9 Lipid Productivity

The lipid content was determined, and the lipid production was calculated based on the results. For both strains, all ratios show increment over time.

Ratio 95:05 leads the production with the highest productivity from day 0 till day 20. As discussed earlier, the lipid content depends on the nitrogen limitation.

Insert (Figure 3.12a and b lipid productivity of *Chlorella vulgaris* and *Chlorella pyrenoidosa* versus time)

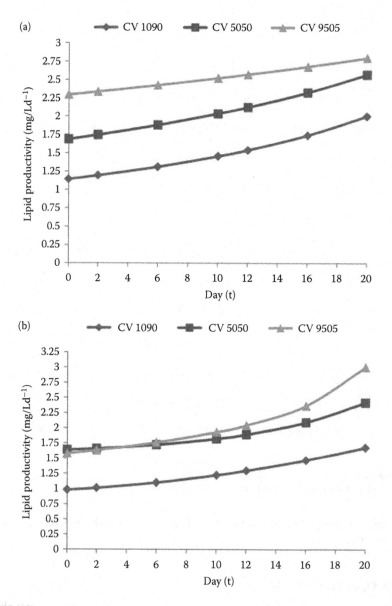

FIGURE 3.12
(a and b) Lipid productivity of *Chlorella vulgaris* and *Chlorella pyrenoidosa* versus time.

3.5 Conclusion

Based on the results obtained from the experiments and analysis, some conclusions can be drawn:

1. *Chlorella pyrenoidosa* was found to be dominant, and the POME condition is favorable for *Chlorella pyrenoidosa*.
2. *Chlorella vulgaris* can adapt and grow well in a POME environment.
3. Optimization of lipid content is best at ratio 95:05 and *Chlorella pyrenoidosa* produces higher lipid content compared to *Chlorella vulgaris*.

3.6 Recommendations

Some recommendations are suggested for future research in this area:

1. Lipid content can be optimized by varying photo light duration from the best strain and nutrient ratio obtained from this study.
2. Lipid production in heterotrophic condition can be investigated using the optimized nutrient ratio.

References

Bersanti, L., Gualtieri, P. 2006. *Algae: Anatomy, Biochemistry and Biotechnology*. FL: CRC Press, Taylor & Francis Group.

Feng, Y., Li, C., Zhang, D. 2011. Lipid production of *Chlorella vulgaris* cultured in artificial wastewater medium. *Bioresource Technology* 102: 101–105.

Fulton, L. 2004. Biomass and Agriculture: Sustainability, markets and policies. International Energy Agency (IEA) biofuels study–Interim report: Result and key messages so far. International Energy Agency, France, pp. 105–112.

Lam, M. K., Lee, K. T., Mohamed, A. R. 2009. Life cycle assessment for the production of biodiesel: A case study in Malaysia for palm oil versus jatropha oil. *Biofuels, Bioproducts and Biorefining*, 3(6): 601–612.

Lang, X., Dalai, A. K., Bakhshi, N. N., Reaney, M. J., Hertz, P. B. 2001. Preparation and characterization of bio-diesel from various bio-oils. *Bioresource Technology* 80: 53.

Lee, J.-Y., Yoo, C., Jun, S.-Y., Ahn, C.-Y., Oh, H.-M. 2010. Comparison of several methods for effective lipid extraction from microalgae. *Bioresource Technology* 101 (suppl. 1): S75–S77.

Liu, Z. Y., Wang, G. C., and Zhou, B. C. 2008. Effect of iron on growth and lipid accumulation in *Chlorella vulgaris*. *Bioresource Technology* 99(11): 4717–4722.

Mata, T. M., Martins, A. A., Caetano, N. S. 2010. Microalgae for biodiesel production and other applications: A review. *Renewable and Sustainable Energy Reviews* 14: 217–232.

MPOB. 2010. Overview of the Malaysian Oil Palm Industry. 2009.econ.mpob.gov.my/economy/Overview_2009.pdf

Neoh, C. H., Lam, C. Y., Yahya, A., Ware, I., Ibrahim, Z. 2015. Utilization of agro-industrial residues from palm oil industry for production of lignocellulolytic enzymes by *Curvularia clavata*. *Waste and Biomass Valorization* 6(3): 385–390.

Parthasarathy, S., Mohammed, R. R., Fong, C. M., Gomes, R. L., Manickam, S. 2016. A novel hybrid approach of activated carbon and ultrasound cavitation for the intensification of palm oil mill effluent (POME) polishing. *Journal of Cleaner Production* 112: 1218–1226.

Thian, X. Y., Md Din, M. F., Nor Anuar, A., Jamalluddin, H. 2010. A local cultivation of microalgae in autotrophic and heterotrophic condition. *Proceedings of Postgraduate Seminar on Water Sustainability (UTM & UNESCO –IHE)*. January 20, 2010. UTM, Johor.

Vijayaraghavan, K., Ahmad, D., Abdul Aziz, M. E. 2007. Aerobic treatment of palm oil mill effluent. *Journal of Environmental Management* 82: 24–31.

Widjaja, A., Chien, C. C., Ju, Y. H. 2009. Study of increasing lipid production from fresh water microalgae *Chlorella vulgaris*. *Journal of the Taiwan Institute of Chemical Engineers* 40: 13–20.

Wu, T. Y., Mohammad, A. W., Jahim, J. M., Anuar, N. 2010. Pollution control technologies for the treatment of palm oil mill effluent (POME) through end-of-pipe process. *Journal of Environmental Management* 91: 1467–1490.

Section II

MBR Technologies

4

Removal of Micro-Pollutants from Wastewater through MBR Technologies: A Case Study on Spent Caustic Wastewater

Noor Sabrina Ahmad Mutamim and Zainura Zainon Noor

CONTENTS

4.1 Introduction

The membrane bioreactor (MBR) is not new in wastewater technology. It has been used and explored since the 1960s in solid–liquid separation where the bioreactor acts as a biological treatment process and the membrane is used as the filter in the filtration process. Generally, the biological process shows greater performance than the filtration process (Widjaja et al., 2010). The membrane plays a role in separating solid and liquid whereby the biological process by activated sludge converts the particle waste into flocs before it is separated by the membrane. In the 1990s, submerged MBR was commercialized and it was found that this system had a lower operational cost (Le-Clech et al., 2006) than any other type of MBR.

Membrane fouling becomes a major factor since membrane is used. Apart from that, high cost of maintenance and operation is needed to maintain the performance of MBR and most of the treatment plants avoid using MBR because of this problem. Other concerns are the limitation in pH, temperature, pressure, and also some corrosive chemicals. If not, it not only the microbes in the reactor are contaminated but the membrane is also destroyed. However, through some research, the fouling factor can be reduced. Some modifications and integration of MBR are done to minimize the constraints. According to the environment and its regulations, this technology is actively manipulated and many new MBR types exist to ensure that the treated water meets the standard.

MBRs have been installed worldwide and the most common manufacturers are Kubota from Japan, Zenon from Canada, and Mitsubishi. Zenon has the technology in treating wastewater which is four times better in performance compared to Kubota (Radjenovic et al., 2008). The challenges while handling MBR are to maintain the solids retention time (SRT) as low as practicable (10–20 days) and to reach optimum permeate flux through fouling control with less energy consumption (Le-Clech et al., 2006).

MBRs have their own characteristics that industries with high micro-pollutants wastewater can choose from. MBRs can give high performance in treating water besides having a smaller footprint compared to conventional activated sludge where the secondary clarifier and tertiary filtration processes are eliminated. MBR also produces high-quality effluent (Chang et al., 2006), is good at removing organic and inorganic contaminants, and is capable of resisting high organic loading (Zhang and Verstraete, 2002), lesser sludge generation (Le-Clech et al., 2006), and easy to adjust SRT. Compared to conventional activated sludge plant (CASP) where the sedimentation process depends on the factor of gravity, it is different with MBR where it can be operated with high mixed liquor suspended solids (MLSS) concentration. Due to the advantages of MBR, some industries will reuse the treated water for other processes, for instance, heating and cooling processes or selling the treated water to other industries. Therefore, the treated water must have a smaller amount of contaminants to avoid sensitive equipment or pipes from breaking down (Radjenovic et al., 2008). The end of this chapter will discuss the treatment of spent caustic wastewater performance by using an aerobic submerged membrane bioreactor (ASMBR).

4.2 General MBR Design

Generally, the basic MBR is divided into two types: the side-stream MBR system and aerobic and anaerobic MBR. For the side-stream MBR, the feed of the wastewater is directly in contact with the biomass. Then, both wastewater and biomass are pumped through the recirculation loop consisting of membranes. The concentrated sludge is then recycled to the reactor while the water effluent is discharged. The purpose of separating the membrane and bioreactor is actually to reduce the power used by the air diffuser (Frederickson, 2005; Sombatsompop, 2007). The immersed system has less operational cost because there is no recirculation loop compared to the side-stream system. A biological process occurs around the membrane. Both need to pump out the excess sludge to maintain sludge age. The pressure driven across the membrane on these systems can either be by suction of the membrane or by applying pressure to the bioreactor. Air bubbles are supplied to both systems for aeration besides scouring especially for the immersed system in reducing membrane fouling (Chang et al., 2006). On the other hand, aerobic and anaerobic MBRs is where oxygen acts as an important medium for microbial growth in the aerobic process whereas the anaerobic process is carried out without oxygen. The anaerobic process is less efficient in removing chemical oxygen demand (COD) and takes a long time to start up. Usually anaerobic treatment is used for treating high strength wastewater at low temperature which is suitable for microbial growth. It is difficult to adjust low temperature for the waste feed as it causes high fouling compared to aerobic at low flux (Judd, 2006).

In CASP treatment, large clarifying basins are needed to make sure the flocs are completely settled but by using membranes, there is no more settling process needed and the area used can be reduced as the membrane acts as a separator (Ng and Kim, 2007). The type of membranes used are also different depending on the size of contaminants during the treatment process. Basically contaminants with the size of a particle from 0.1 to 10 μm use microfiltration for removing suspended particles, ultrafiltration (UF) for particle size 1000–5000 dalton for instance in bacteria and viruses, and nanofiltration (NF) for particles with 100–1000 dalton in molecular weight. In treating high strength wastewater with shock loading of matter, microfiltration is chosen among the others to prolong membrane usage. According to Judd (2006), all types of membranes lie in pressure-driven process. Most of the treatment plants use microfiltration or UF instead of NF due to fouling and cost factors (Judd, 2006; Le-Clech et al., 2006; Radjenovic et al., 2008; Sombatsompop, 2007).

The two types of materials used to construct membranes are polymeric and ceramic. Suitable polymers that have been used commercially include polyvinylidene difluoride (PVDF), polyethylsulfone (PES), polyethylene (PE), and polypropylene (PP) because of good physical and chemical resistance but PE is more quickly fouled compared to PVDF (Le-Clech et al., 2006). Most MBR use microfiltration sizes to reduce the level of fouling (Marti et al., 2011). Membrane configurations also play an important part whereby for each configuration, they have their own advantages and disadvantages based on the cost, capability to withstand turbulence and be backflush-able (Judd, 2006). In the 1960s, membrane was considered a costly treatment because the material used to set up and maintain the membrane was expensive besides being difficult to find. The cost of membrane material decreased with time and technology growth. Therefore, the cost of membrane also declined.

4.3 General MBR Operation Parameters

CASP has been used for a long time. However, when related with high strength wastewater, this method will not be able to cope with the high content of organic loading and inorganic matter because of low biodegradability and inhibitor and in some cases (Lin and Ho, 1996), the microbes can be destroyed because of shock loading of matter. This is because microbes take a long time to biodegrade the matter and a high concentration of microbes is needed to ensure that all the organics are totally biodegradable. CASP cannot tolerate high concentration of microbes due to low efficiency of aeration. Basically, SRT is operated coupled with HRT for CASP.

SRT is the solids or flocs growth in many sizes that needs to be retained in the plant before it settles down for a period of time whereas HRT is the time taken for matter to pass through the plant. This means that CASP relies on both to make sure the flocs are really settled before going on to other treatments (Judd, 2006). Also, CASP needs microbes with fast growth and flocs formation species. Otherwise if the microbial growth rate is low, it will lead to washing out together with the excess 246 sludge because of the shortage of SRT. Consequently, the production of sludge and F/M ratio (food/microorganism ratio) is high and ends up with high sludge excess and disposal covering around 50%–60% of the total cost of wastewater treatment (Radjenovic et al., 2008). To get the best performance in treating high strength wastewater, the concentration of sludge must be high to increase the process of degradation.

The problem when treating wastewater using MBR is that the increased concentration of sludge hastens membrane fouling (Melin et al., 2006). To avoid this problem, the transmembrane pressure (TMP) and flux need to be observed. The best performance for those parameters needs to be identified. It becomes a benchmark in measuring the resistance of membrane, driving force for each unit membrane area, and nearly fouling and the time for cleaning the membrane (Judd, 2006). When flux increases, TMP also increases. It shows that there is more wastewater that can be separated until the TMP drops when the flux continues to increase. The decreasing phase shows that membranes have a high resistance and need cleaning before the membrane becomes fouled leading to membrane damage.

On the other hand, in the MBR system, SRT and HRT do not rely on each other because MBR is more focused on membrane filtration rather than settling by gravity and this system does not consider the flocs growth but still maintains the minimum sludge production with low F/M ratio (less substrate is present per unit of biomass) (Ng and Kim, 2007; Radjenovic et al., 2008) and retains the microorganisms in the reactor and the sludge age (Benedek and Côté, 2003; Judd, 2006; Noor et al., 2002; Widjaja et al., 2010). Besides that, the formation of flocs makes it easier to filter. However, if the F/M is too low, microorganisms in the activated sludge will not grow well (Widjaja et al., 2010), or else if MBR has a very high MLSS, it will lead to clogging, low efficiency of aeration, and the need for a large bioreactor (increases the initial capital cost) (Radjenovic et al., 2008). HRT with low levels will increase the organic loading rate (OLR) which will end up with reactor volume reduction and reduce the performance of MBR whereas if the HRT is high, MBR has a good performance (Fallah et al., 2010; Jianga et al., 2008; Viero and Ferreira Jr., 2008).

An increase in the SRT will decrease the soluble microbial product (SMP), whereby the microorganisms will stay longer in the reactor and prolong the biological degradation process. However, if the SRT is lower, it will increase the level of the SMP and fouling (Jianga et al., 2008; Liang et al., 2007). Nevertheless, if the SRT is too long, it tends to foul the membrane with the accumulation of matter and high viscosity of sludge (Jianga et al., 2008). With an anaerobic bioreactor, HRT and SRT are independent but dependent on biomass retention. It also produces methane as a side product, and odor, (Judd, 2006) and does not use any aeration process and is energy saving, as methane can be collected for energy generation (Judd, 2008).

Hai et al. (2008) applied hollow fiber membrane with 50–200 μm pore of membrane for treating high strength wastewater from textile industries consisting of various chemicals, organic loading, and color. The color and total organic compound (TOC) removal is between 93% and 97%. Other researches show high concentration of ammonium (10,500 mg L^{-1} of COD and 1220 mg L^{-1} of NH_4^+) treated by anaerobic and aerobic MBRs which give a good result when the percentage of removal of COD is almost 99% and 0.5 g $NH4^+$-N L d^{-1} where denitrified of ammonium is exceeded from 84% to 94% in aerobic condition.

In anaerobic MBR treatment, only 60%–80% of COD removal and methane was produced in a huge amount (Zhang and Verstraete, 2002). For different HRTs between 5 and 10 days, the result showed COD removal of around 78%–96%, 87%–99% for BOD5 removal, and 92%–95% oil and grease removal by using aerobic MBR at 40°C (mesophilic–thermophilic transitional temperature) (Kurian et al., 2006). Ahn et al. (2007) showed that the integrated anaerobic upflow bed filter (AUBF) and aerobic MBR gave 99% of organic removal and 46% of total nitrogen removal for strong nitrogen wastewater. Widjaja et al. (2010) reported that by using a submerged MBR, the percentage removal of 1500 and 2500 mg L^{-1} of COD (F/M is 0.2–0.6 kg COD kg^{-1} MLSS.day^{-1}) and MLSS 8000 mg L^{-1} with PAC added is between 86.69% and 91.86%. Viero et al. (Viero and Ferreira Jr., 2008 and Viero et al., 2008) showed a comparison with HRT is 12 h, the percentage of phenol removal is 94% by using a fixed-biofilm reactor (Hsien and Lin, 2005) but by using CASP, the percentage is around 61% (Lu et al., 2006) and 100% removal for high loading of synthetic phenol wastewater by using an external MBR (Vazquez et al., 2006).

Kurian et al. (2006) stated that by increasing the HRT from 5 days to 10 days, the oil and grease removal also increased from 92% to 95%; the same goes for COD and BOD where the percentage removal increased from 78% to 97% for COD and 87% to 99% for BOD. However if the HRT is too high, it will enhance the cost of operation through maintenance and energy costs. Shin et al. (2005) reported that integrating the submerged MBR with an AUBF reactor in the range of pH 7.92–8.57 is due to the high content of ammonia. The challenging part of this treatment is when the 183 nonbiodegradable matter content is too high but the percentage of COD removal is about 91% and for nitrification it is 60%. This system generates organic acid for denitrification. However, when it reaches the stage where acid is more than the denitrification needed, it produces acidic wastewater that needs adjustment before entering the submerged MBR. Otherwise it causes difficulty in adaptation for the sludge in the submerged MBR and the worst consequence is it will reduce the performance of the microbial process.

4.4 Case Study: ASMBR in Treating Spent Caustic Wastewater

Most countries which produce petroleum today generate a lot of income because of its high worldwide demand. According to the Energy Information Administration, Malaysia has six refineries with a total capacity of 554,832 barrels per day (bbl d^{-1}) and the three largest refineries include the Shell Port Dickson Refinery, Petronas Melaka-I, and Petronas Melaka-II with 155,000, 92,832, and 126,000 bbl d^{-1}, respectively (EIA, 2010). With the increasing capacity of petroleum production, the level of pollution to the environment also increases. Now, the government takes the initiative to reduce and cut down the level of pollution by inviting research done specifically on this issue.

Caustic soda, also known as sodium hydroxide is widely used in industries as the main ingredient for soap and detergents. It is used in making soft drinks and glass, in the chemical peeling of fruits, and for many other purposes. For instance, caustic soda is used as a detergent for cleaning drums or tanks in many industries, especially in petrochemical industries because of its alkalinity value in removing obstinate dirt. Besides that, it is very essential as a raw material in pulping and bleaching processes. Today's refineries and petrochemical producers are forced by more stringent environmental regulations to better monitor and control their wastewater. The spent caustic from refineries and petrochemical plants is categorized in the Environmental Quality Act 1974 under the First schedule of Environmental Quality Act 2005 (Scheduled Wastes) Regulation 2005 (SW 314-Oil or sludge from oil refinery or petrochemical plant) (Environmental Quality Act, 1974). The spent caustic should be treated to comply with the standard where the recovered, recycled, and reconstituted processes should achieve the standard and specification categorized as scheduled waste.

One of the key contributors to relatively high COD and biological oxygen demand (BOD) is from the acid gas both CO_2 and hydrogen sulfide (H_2S) removal systems. H_2S is known as a corrosive element that will cause problems in the refinery process. Therefore, they manage to use caustic soda as a material in improving the quality of the product. Typically, they used dilute caustic soda (such as NaOH) as the active reagent in the removal system. Here, caustic soda also plays a role in removing the smell coming from H_2S and neutralizes the acidic materials from hydrocarbon and offgases. The result of the washing for refinery process or scrubbing during ethylene process, the waste stream is producing spent caustic (Gondolfe and Kurukchi, 1997).

Typical refinery spent caustic derived from acid gas removal either in a petrochemical or refinery facility can contain significant contaminants comprised of dissolved hydrocarbon, polymers, and active polymer precursors in addition to well-defined levels of sodium salts and free caustic. These contaminants can grossly inhibit the conversion level of sulfides (S^{2-}) to its highest oxidation state of sulfates, SO_4^{2-}. Spent caustic solution may be corrosive to the skin, eyes, and respiratory tract and may be harmful if swallowed, inhaled, or absorbed through the skin if the solution is overdosed. Spent caustic solution is also expected to be an eye and skin irritant. It can also be a respiratory irritant if there is exposure to high vapor levels. Lung damage may occur if swallowed and aspirated. There are no known chronic, mutagenic, or teratogenic hazards. It is also known as to have high ecotoxicity to marine life according to pH and corrosivity characteristics (American Petroleum Institute, 2009).

Figure 4.1 shows the stages of refined wastewater treatment. In general, physicochemical processes are used to treat spent caustic but these consume high operational cost and also cause hazardous environmental impact due to the use of high temperature, pressure, and chemical. The organic and inorganic substances in spent caustic are also partly oxidized in the physicochemical processes and still need further treatments. Spent caustic is commercially treated by applying wet air oxidation, liquid incineration, or disposal by using deep-well injection (Sipma et al., 2004). Treatment is commonly continued with biological treatment because this has the ability to enhance the removal of organic and inorganic pollutants. Sulfides in spent caustic treated by a bioreactor may be biologically and chemically oxidized into sulfate (Gerardi, 2006). Sulfides in spent caustic oxidize biologically and produce sulfate as a result of complete oxidation and sulfur under limited oxygen and the reactions are shown in Equations 4.1 and 4.2 below (Lohwacharin and Annachhatre, 2010; Sipma et al., 2004):

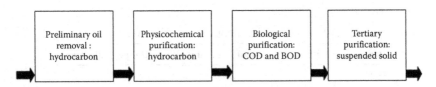

FIGURE 4.1
General refinery wastewater treatment. (Adapted from Berne, F. and Cordonnier, J. 1995. *Industrial Water Treatment: Refining, Petrochemicals and Gas Processing Techniques.* Gulf Publishing Company: Houston, Texas, USA.)

$$HS^- + 2O_2 \rightarrow SO_4^{2-} + H^+ \quad \Delta G = -210.81 \text{ kJmol}^{-1} \tag{4.1}$$

$$HS^- + 0.5O_2 \rightarrow S^\circ + OH^- \quad \Delta G = -796.48 \text{ kJmol}^{-1} \tag{4.2}$$

The focus on the ASMBR is because of the efficiency in removing the contaminants and the production of high-quality effluent. Previous experiments have shown the ASMBR effectiveness in removing COD and color (Yuniarto et al., 2008). The effectiveness of the ASMBR will also depend on the types of membrane used such as flat sheet, hollow fiber, tubular, and spiral wound. All types of membranes chosen depend on the types of wastes that need to be treated. The sizes of the membrane pores also play an important role to create the best ASMBR. It depends on the particles in the wastewater that need to be removed. Table 4.1 shows the characteristics of spent caustic wastewater. Spent caustic was designed according to Sipma et al. (2004). The spent caustic wastewater pH reading is 10 ± 1 and needs to be neutralized before it can be treated by MBR. Sulfuric acid is used as pH adjustment and below 10% of spent caustic component removal. Nonsulfide is hard to oxidize like phenol due to its high stability component. Sulfide is known as an oxidized scavenger and is easy to oxide by pH adjustment (Mara and Horan, 2003; Sheu and Weng, 2000).

The schematic diagram for the laboratory scale ASMBR is shown in Figure 4.2. The specification of membrane is given in Table 4.2. The experiments were carried out at room temperature of $26°C \pm 1$. The U-shaped hollow fiber membrane was vertically submerged into the 4 L bioreactor and flowed using outside-in mode. Air supply was controlled at 15 L min^{-1} at the bottom of the membrane unit for oxygen supply to the biomass and to create turbulence which helped in reducing membrane fouling. The feed entering the bioreactor was controlled by using a water leveler. The feed and effluent were assumed to have the same flow rate. The effluent was drawn from the membrane by using a peristaltic pump. The pH of mixed liquor was maintained at 7–7.5. The operation conditions and parameters are shown in Table 4.3. During the studies, the fouled membranes were removed by backflush; in which the backflush flux doubled from the filtration flux

TABLE 4.1

Characteristic of Synthetic Spent Caustic Wastewater

Parameter (mg L⁻¹)	Before pH Adjustment	After pH Adjustment
COD	2300	2120
BOD	810	754
Sulfide	102	88
Sulfate	83	403
Phenol	48	44

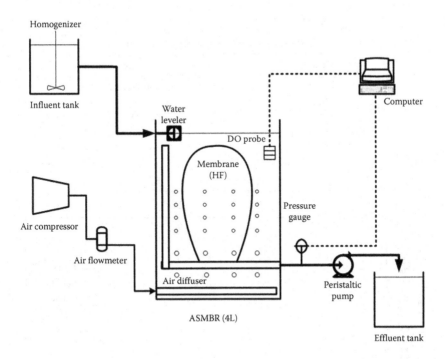

FIGURE 4.2
Flow diagram of aerobic ASMBR.

TABLE 4.2

Membrane Specification for ASMBR

Characteristic	Description
Membrane material	Polyethersulfone
Membrane configuration	U-shaped hollow fiber
Maximum filtration	70 kPa
Pore size, Ø	0.2–0.02 μm
Surface area per module, A	0.075 m²
Number of fibers per module	100
pH	2–13

at duration of 60 seconds for reversible fouling and chemical cleaning by 0.5% ppm NaOCl for 24 h, if irreversible fouling had occurred.

In this case study, the biomass was acclimatized until steady-state was achieved. Then, the steady-state biomass was transferred into the MBR and the biomass was acclimatized again for adaptation in the MBR's new environment. Figure 4.3 shows the COD, sulfide and sulfate effluent concentration for a 10-day operation. The result reports 87% of COD in the beginning and it increases up to 99% as the MLSS accumulation in the

TABLE 4.3

Operation Condition of Lab-Scale ASMBR

Parameter	Value
MLSS (mg L^{-1})	5000 ± 23
Flux (LMH)	4
pH	7.2–7.8
DO (mg L^{-1})	2–7
HRT (hours)	16
SRT (days)	40
Permeate/relax (min)	5/1

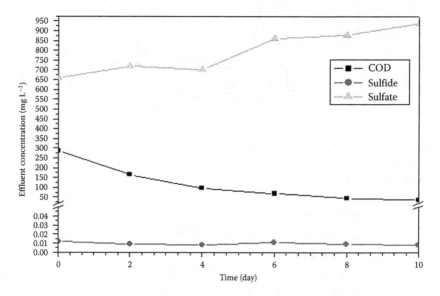

FIGURE 4.3
Effluent concentration of ASMBR treatment.

reactor increases with time. Sulfide was almost totally removed in the 10-day operation and sulfate increases from 36% to 52% with F/M ratio of the 0.633 kg COD kg MLSS^{-1} d^{-1}. Effluent concentration for phenol was recorded 0.012 mg L^{-1} and acceptable for water discharge. Figure 4.4 shows TMP performance and a cake layer form on the surface on the membrane. TMP increases gradually due to formation of the biocake layer on the membrane surface and this biocake layer becomes a "secondary membrane." The form of biocake layer reduced the lifespan of the ASMBR operation due to membrane fouling. Membrane fouling also occurs by SMP and EPS influence. SMP and EPS are biomass products during cell lysis and are able to attach onto membrane surface and attach or discharge through membrane pores (Yuniarto et al., 2013).

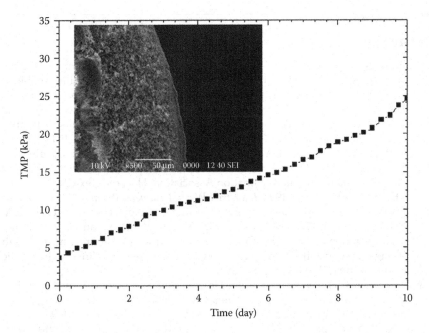

FIGURE 4.4
TMP of ASMBR.

4.5 Conclusion

Most high strength spent caustic wastewater containing organic and inorganic sulfides are successfully treated by MBR especially in COD removing. By controlling the parameters such as SRT, HRT, TMP, flux, and MLSS to optimum conditions the best MBR performance can be produced. The wastewater needs to be neutralized before entering the MBR to avoid short life of biomass. However, fouling the membrane becomes a major barrier in MBR treatment. There are also some methods to reduce fouling problems and enhance the performance of MBRs and produce high-quality effluents. This technology can grow and many researchers are aggressively modifying the MBR system based on wastewater conditions in order to increase the performance and at the same time reduce operational cost. The membrane is the heart of this system and it is also sensitive to unusual chemicals besides pH, pressure, and temperature that contribute to reducing the performance of MBR. This problem needs to be of concern since the condition of wastewater change depends on the processes and the environment involved. Sometimes, modifications of wastewater such as dilution for high strength and toxic wastewater or neutralization of acidic or base of wastewater need to be done to prolong the lifespan of the membrane besides maintaining microbial growth.

References

Ahn, Y. T., Kang, S. T., Chae, S. R., Lee, C. Y., Bae, B. U., and Shin, H. S. 2007. Simultaneous high-strength organic and nitrogen removal with combined anaerobic upflow bed filter and aerobic membrane bioreactor. *Desalination*, 202, 114–121.

American Petroleum Institute, A. 2009. Composition of spent sulfidic caustic streams. In *Acids and Caustics from Petroleum Refining Category*. U.S.E.H.C.P. Petroleum HPV Testing Group (Ed.). American Petroleum Institute: Washington, pp. 3–28.

Benedek, A. and Côté, P. 2003. *Long Term Experience with Hollow Fibre Membrane Bioreactor*. International Desalination Association: Houston, Texas, USA.

Berne, F. and Cordonnier, J. 1995. *Industrial Water Treatment: Refining, Petrochemicals and Gas Processing Techniques*. Gulf Publishing Company: Houston, Texas, USA.

Chang, J.-S., Chang, C.-Y., Chen, A.-C., Erdei, L., and Vigneswaran, S. 2006. Long-term operation of submerged membrane bioreactor for the treatment of high strength acrylonitrile–butadiene–styrene (ABS) wastewater: Effect of hydraulic retention time. *Desalination*, 191, 45–51.

EIA, E. I. A. 2010. Country Analysis Briefs Header-Malaysia. Retrieved 4th September, 2011, from http://205.254.135.24/emeu/cabs/Malaysia/Full.html

Environmental Quality (Sewage and Industrial Effluents) Regulations 1974.

Fallah, N., Bonakdarpour, B., Nasernejad, B., and Moghadam, M. R. A. 2010. Long-term operation of submerged membrane bioreactor (MBR) for the treatment of synthetic wastewater containing styrene as volatile organic compound (VOC): Effect of hydraulic retention time (HRT). *Journal of Hazardous Materials*, 178, 718–724.

Frederickson, K. C. 2005. *The Application of a Membrane Bioreactor for Wastewater Treatment on a Northern Manitoban Aboriginal Community*. Master of Science, University of Manitoba: Winnipeg, Manitoba, Canada.

Gerardi, M. H. 2006. Nitrogen, phosphorus and sulfur bacteria. In *Wastewater Bacteria*. M. H. Gerardi (Ed.). Wiley: Williamsport, Pennsylvania.

Gondolfe, J. M. and Kurukchi, S. A. 1997. *Spent Caustic Treatment: The Merits of Pretreatment Technology*. Shaw Stone & Webster (patent pending).

Hai, F. I., Yamamoto, K., Fukushi, K., and Nakajima, F. 2008. Fouling resistant compact hollow-fiber module with spacer for submerged membrane bioreactor treating high strength industrial wastewater. *Journal of Membrane Science*, 317, 34–42.

Hsien, T.-Y. and Lin, Y.-H. 2005. Biodegradation of phenolic wastewater in a fixed biofilm reactor. *Biochemical Engineering Journal*, 27, 95–103.

Jianga, T., Myngheer, S., Pauw, D. J. W. D., Spanjers, H., Nopens, I., Kennedy, M. D. et al. 2008. Modelling the production and degradation of soluble microbial products (SMP) in membrane bioreactors (MBR). *Water Research*, 42, 4955–4964.

Judd, S. 2006. *Principles and Applications of Membrane Bioreactors in Water and Wastewater Treatment* (1st ed.). Elsevier: UK.

Judd, S. 2008. The status of membrane bioreactor technology. *Trends in Biotechnology*, 26(2), 109–116. Retrieved from.

Kurian, R., Nakhla, G., and Bassi, A. 2006. Biodegradation kinetics of high strength oily pet food wastewater in a membrane-coupled bioreactor (MBR). *Chemosphere*, 65, 1204–1211.

Le-Clech, P., Chen, V., and Fane, T. A. G. 2006. Fouling in membrane bioreactors used in wastewater treatment. *Journal of Membrane Science*, 284, 17–53.

Liang, S., Liu, C., and Song, L. 2007. Soluble microbial products in membrane bioreactor operation: Behaviors, characteristics, and fouling potential. *Water Research*, 41, 95–101.

Lin, S. H. and Ho, S. J. 1996. Catalytic wet air oxidation of high strength industrial wastewater. *Applied Catalysis B: Environmental*, 9, 133–147.

Lohwacharin, J. and Annachhatre, A. P. 2010. Biological sulfide oxidation in an airlift bioreactor. *Bioresource Technology*, 101, 2114–2120.

Lu, J., Wang, X., Shan, B., Li, X., and Wang, W. 2006. Analysis of chemical compositions contributable to chemical oxygen demand (COD) of oilfield produced water. *Chemosphere*, 62, 322–331.

Mara, D. and Horan, N. 2003. *The Handbook of Water and Wastewater Microbiology*. D. Mara and N. Horan (Eds.). *Microbiology of Wastewater Treatment*. Elsevier: UK, p. 459.

Marti, E., Monclús, H., Jofre, J., Rodriguez-Roda, I., Comas, J., and Balcázar, J. L. 2011. Removal of microbial indicators from municipal wastewater by a membrane bioreactor (MBR). *Bioresource Technology*, 102(8), 5004–5009.

Melin, T., Jefferson, B., Bixio, D., Thoeye, C., Wilde, W. D., Koning, J. D. et al. 2006. Membrane bioreactor technology for wastewater treatment and reuse. *Desalination*, 187(1–3), 271–282.

Ng, A. N. L. and Kim, A. S. 2007. A mini-review of modeling studies on membrane bioreactor (MBR) treatment for municipal wastewaters. *Desalination*, 212, 261–281.

Noor, M. J. M. M., Nagaoka, H., and Aya, H. 2002. Treatment of high strength industrial wastewater using extended aeration-immersed microfiltration (EAM) process. *Desalination*, 149, 179–183.

Radjenovic, J., Matosic, M., Mijatovic, I., Petrovic, M., and Barceló, D. 2008. Membrane bioreactor (MBR) as an advanced wastewater treatment technology. *The Handbook of Environmental Chemistry*, 5, 37–101.

Sheu, S.-H. and Weng, H.-S. 2000. Treatment of olefin plant spent caustic by combination of neutralization and Fenton reaction. *Water Research*, 35, 2017–2021.

Shin, J.-H., Lee, S.-M., Jung, J.-Y., Chung, Y.-C., and Noh, S.-H. 2005. Enhanced COD and nitrogen removals for the treatment of swine wastewater by combining submerged membrane bioreactor (MBR) and anaerobic upflow bed filter (AUBF) reactor. *Process Biochemistry*, 40, 3769–3776.

Sipma, J., Svitelskaya, A., Mark, B. V. D., Pol, L. W. H., Lettinga, G., Buisman, C. J. N. et al. 2004. Potentials of biological oxidation processes for the treatment of spent sulfidic caustics containing thiols. *Water Research*, 38(20), 4331–4340.

Sombatsompop, K. M. 2007. *Membrane Fouling Studies in Suspended and Attached Growth Membrane Bioreactor Systems*. Asian Institute of Technology School of Environment, Resources & Development Environmental Engineering & Management. Thailand.

Vazquez, I., Rodriguez, J., Maranon, E., Castrillon, L., and Fernandez, Y. 2006. Simultaneous removal of phenol, ammonium and thiocyanate from coke wastewater by aerobic biodegradation. *Journal of Hazardous Materials*, B137, 1773–1780.

Viero, A. F. and Sant'anna, G. L. Jr. 2008. Is hydraulic retention time an essential parameter for MBR performance? *Journal of Hazardous Materials*, 150, 185–186.

Viero, A. F., Melo, T. M. d., Torres, A. P. R. Ferreira, N. R. Sant'anna, G. L. Jr., Borges, C. P. et al. 2008. The effects of long-term feeding of high organic loading in a submerged membrane bioreactor treating oil refinery wastewater. *Journal of Membrane Science*, 319, 223–230.

Widjaja, T., Soeprijanto, and Altway, A. 2010. Effect of powdered activated carbon addition on a submerged membrane adsorption hybrid bioreactor with shock loading of a toxic compound. *Journal of Mathematics and Technology*, 3, 139–146.

Yuniarto, A., Noor, Z. Z., Ujang, Z., Olsson, G., Aris, A., and Hadibarata, T. 2013. Bio-fouling reducers for improving the performance of an aerobic submerged membrane bioreactor treating palm oil mill effluent. *Desalination*, 316, 146–153.

Yuniarto, A., Ujang, Z., and Noor, Z. Z. 2008. Performance of bio-fouling reducers in aerobic submerged membrane bioreactor for palm oil mill effluent treatment. *Journal Teknologi UTM*, 49, 555–566.

Zhang, D. and Verstraete, W. 2002. The treatment of high strength wastewater containing high concentrations of ammonium in a staged anaerobic and aerobic membrane bioreactor. *Journal Environmental Engineering Science*, 1, 303–310.

5

The Outlook on Future MBR Technologies

**Rabialtu Sulihah Ibrahim, Adhi Yuniarto, and
Siti Nurhayati Kamaruddin**

CONTENTS

5.1 Introduction

Recently, an inventive technology in wastewater treatment called membrane bioreactor (MBR) technology has been extensively employed all over the world. This technology is a biological process that modifies secondary and tertiary treatments employing a membrane filtration process. MBR is able to carry out great effluent achievement with low turbidity, less total bacterial content, and low total suspended solid (TSS) as well as nephelometric turbidity unit (NTU) because of membranes which are used to set up necessary solid–liquid separation. MBRs are applied in treating biologically active wastewater occurring from municipal or industrial sources. This technology is also against biological treatments like the conventional activated sludge (CAS) process, sequence batch reactor (SBR), and others which are used to treat several types of wastewater. Generally, in wastewater application, membranes are applied when an advanced treatment standard is necessary compared to the conventional process. Moreover, MBR is usually cost effective for reuse compared to conventional process that is not normally for reuse.

MBR is a system which integrates the activated sludge system and membrane filtration. In this MBR system, the sedimentation process is taken over by the filtration by membrane. The biological unit in an activated sludge system is responsible for the biodegradation of waste compounds while the membrane module plays a role for physical separation of the treated water from the mixed liquor (Hoinkis et al., 2012). The biological process in MBR changes dissolves organic matter into suspended biomass, decreases membrane fouling, and could increase in recovery (Friha et al., 2014). MBR is suitable in treating several types of domestic and industrial wastewater. MBR can compromise with either composition of wastewater or variation of wastewater flow rate as well as can be used in cases where demands on the effluent are larger than the capability of the CAS process.

Besides that, membrane technology is capable in producing a huge improvement since membranes can produce water of exceptional purity that can be recycled for reuse in different places. There are higher demands on water supplies due to high anticipation for water reuse that relies upon lifestyle alterations. There is already inequality in many areas of the world, and without more water restoration, it will be hard to add food supplies and to maintain environmental conveniences (Howell, 2004). Thus, an advance in membrane technology has high potential to cover the water supply problem due to its performance in producing high-effluent quality. In this case, MBR is responsible in dealing with hard waste possibly with outlined organisms in treating special wastes and is applicable in small area.

MBR offers many advantages and has been familiar for some time. Free selection of hydraulic retention time (HRT) and sludge retention time (SRT) are offered by MBR that allow more pliable control of operational parameters. Moreover, MBR also produces larger sludge concentrations in a bioreactor permitting effective treatment of high strength wastewater. Low biodegradable pollutants contained in wastewater are also able to be removed contributing to the removal of the recalcitrant compounds effect by the improvement of specialized and slow-growing microorganisms. This is because the contact time of sludge and critical classes of substrates are due to the retention of activated sludge containing solids and macro molecules combined with long sludge age (Melin et al., 2006).

MBR is widely used around the world in treating various types of wastewater. The application and performances of this technology is generally subject to proper selection of membrane material and operational process which become the drivers toward the extent of implementation (Tiranuntakul, 2012 and Santos et al., 2011). Due to the progress of MBR research, the main limitation in this technology which is membrane fouling could be recovered and full-scale membrane plant operations of MBR technology were reliable in the subsequent wastewater treatment (Tiranuntakul, 2012). Furthermore, Radjenovic et al. (2008) mentioned that in the last 15 years, whole life costs of the membrane decreased more than eight times, which closed the gap in prices between CAS and MBR technologies. Thus, it can be concluded that MBR technology can be highly competitive and its future widespread application should be possible due to expected increased membrane lifetime, sufficient full-scale plants being successfully operated, as well as guaranteed to be reliable.

5.2 Performances and Application of MBR

There are several types of MBR and its application in various types of wastewater treatment. Tiranuntakul (2012) reported that MBR technology is applied mainly for the treatments of heavily loaded wastewater like oily wastewater, wastewater from textile industries, or discharges from tanneries, landfill leachate, chlorinated solvents in manufacturing wastewater, and also groundwater remediation. MBR is used as a secondary treatment in industries in reducing biodegradable and nonbiodegradable matter in the end product or in advanced treatment to remove residual nutrients that are not fully removed during secondary treatment (Tchobanoglous et al., 2004). Table 5.1 shows the comparison performances on several types of MBR with different types of wastewater.

5.3 Advanced Technology of MBR

Advanced MBR technology of MBR has developed due to the presence of several limitations. There are several advanced MBRs being studied in order to achieve good effluent quality, prolonged operational time and less energy, and economical consumption. Liu et al. (2013) studied a new MBR combined with an added bio-electrochemical cell containing iron anodes, microbes, and conductive membrane cathodes altered by polypyrrole, that keep the advantages of MBR and produced fixed electrical potential for cathode membrane fouling reduction. The biofilm microbes on the iron anode can abstract electrons from wastewater like a microbial fuel cell (MFC), and the biochemical and chemical corrosion of the iron anode also created electrons as a chemical cell. This type of MBR was developed since the increasing external energy expenditure in aeration for membrane fouling reduction leads to limiting the employment of MBR in wastewater treatment. The general energy consumption could be lowered without adversely affecting wastewater treatments if the energy from wastewater in MBR was applied directly to lessen fouling.

In this research, Figure 5.1 shows the mechanism and schematic diagram of bio-electrochemical membrane reactor (BEMR). When analyzing with control MBR, the membrane fouling in BEMR had been decreased significantly. TMP changes prove the important declining of fouling in BEMR. Alleviation of fouling occurred due to electrostatic repulsion by electrons and flocculation by Fe^{3+} from anode corrosion. The back diffusion of foulants (colloids and sludge) might improve and filter cake should lessen; the specific resistance of sludge might reduce resulting from flocculation by Fe^{3+} because of the repulsion force (Liu et al., 2012). In this BEMR, without any input of external electric energy, membrane fouling was well fixed by the

TABLE 5.1

Performances of Several Types of MBR

Type of MBR	Type of Wastewater	Comments
Submerged MBR; Friha et al. (2014)	Cosmetic wastewater	98.13% and 83.73% of COD at OLR of 1500 mg COD/m^3.d and HRT of 27 h were removed Biomass concentration inside the MBR was lower compared to the conventional process MLSS declines with the increase of OLR due to the accumulation of nonbiodegradable organic compounds inside the bioreactor
Aerobic submerged; Yigit et al. (2009)	Textile industry	Large removal efficiencies are managed to be obtained for several parameters due to high MLSS concentrations A significant degree of denitrification was achieved hence anoxic region is produced among flocs or at the bottom of the reactor The dissolved solids could not be removed by ultrafiltration as employed in MBR processes, further treatment (i.e., nanofiltration or reverse osmosis) to remove dissolved solids is required prior to reuse of MBR effluent, especially in dyeing and finishing processes
Submerged MBR; Katayon et al. (2004)	Food industry wastewater	The mean flux value decreased with the increase of MLSS concentrations At low MLSS and horizontally positioned membrane modules, turbidity and suspended solid removal were quite high
Submerged MBR; Rosenberger et al. (2002)	Municipal wastewater	Anoxic zone was also provided to enable denitrification process The performance of MBR is better than conventional sewage works. At low F/M ratio, COD reduction reached 95% and was extremely stable.
Submerged MBR; Sadri et al. (2008)	Landfill leachate	Excellent performance has been shown by MBR as it managed to give 97% and higher BOD removal, even at low HRT Sufficient suspended solids removal and nitrification of the incoming ammonia as well as important elimination of several metals such as iron, lead, manganese, cadmium, and aluminum from feed of leachate was also observed
Moving bed MBR (MB-MBR); Jamal Khan et al. (2012)	High strength synthetic wastewater	MB-MBR performed higher filtration runs because of less membrane fouling resistance, for reversible and irreversible compared to conventional MBR

(Continued)

TABLE 5.1 (*Continued*)

Performances of Several Types of MBR

Type of MBR	Type of Wastewater	Comments
Attached growth; Sombatsompop et al. (2006)	Synthetic wastewater	The result of TKN (total Kjehldahl nitrogen elimination performances in the attached growth process was in the range of 83%–90% and was higher than in the suspended growth process. Particle fouling in the attached growth reactor was less than in the suspended growth reactor
Submerged MBR; Huang et al. (2008)	Synthetic domestic wastewater	Good effect: mechanically scouring membrane surface Bad effect: disintegrate sludge flocs Effective carrier dosage mitigated fouling effectively
Submerged MBR; Yang et al. (2006)	Terephthalic acid wastewater	Membrane permeability improved and minimized hydraulic resistance Cake removal influenced by scouring of suspended carriers was the major donator to control membrane fouling.
Attached growth MBR; Achilli et al. (2011)	Synthetic wastewater	Attached growth of MBR process had a 30% larger critical flux value but higher biomass production Attached growth media had an important role in managing membrane fouling

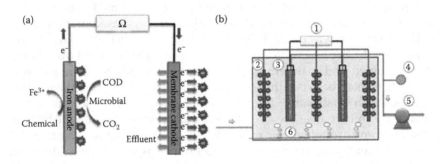

FIGURE 5.1
The mechanism (a) and schematic diagram (b) of BEMR ((1) external resistor, (2) iron anode electrode, (3) membrane cathode module, (4) vacuum meter, (5) peristaltic pump, and (6) aerators). (Adapted from Liu et al., 2013. *Journal of Membrane Science*, 430, 196–202.)

negative charge and flocculating agent from the anode electrode. Compared to anaerobic effluent (such as MFC anodic effluent), BEMR shows better effluent with the pH at around 6–7, the COD and turbidity removal rates were moved to 100% without odor. Thus, there were energy and environmental advantages in BEMR, involving good effluent quality, longer operational time, and half self-supplied energy.

Besides that, in another research project, a novel enhanced MBR by internal micro-electrolysis (IE) was significantly employed for treatment of reactive brilliant blue X-BR dye wastewater due to its larger organic matter and nitrogen removal. According to Qin et al. (2012), this study focuses on investigating the consequence of iron ions discharged from the IE on properties of biomass and membrane filtration. Before aerobic MBR treatment process, the textile wastewater must undergo pretreatment process. So, IE, as one electrochemical oxidation process, has been gaining attention currently, among several physical and chemical processes considered to have minor or no harmful impact on the environment for the use of inoffensive reagents and requiring no excess energy (Cheng et al., 2007).

Figure 5.2 shows the schematic diagram of the process in enhanced MBR by internal micro-electrolysis. In this research, a hybrid MBR (HMBR) with iron ions was fed and an iron controlled MBR (CMBR) was conducted in parallel function to investigate the effects of iron ions liberated from the IE process on the property of the sludge and membrane fouling. In cultivating the degradation of reactive brilliant blue X-BR wastewater, internal electrolysis (IE) is coupled with aerobic MBR. For CMBR, the supernatant was put into the CMBR after adjusting pH to 9.0–10.0 by NaOH addition so that the iron ions of the Fenton effluent will significantly be removed but for HMBR the Fenton effluent was directly moved into the bioreactor. From the observations, a massive number of iron ions were discharged from the galvanic corrosion in the IE pretreatment process. As iron is a significant and crucial component of biomolecules, the performance of activated sludge was improved. Some research claimed that Fe^{3+} was useful in developing the filterability of mixed liquor. Based on Figure 5.2, the IE experiment was operated under the optimal reaction from the previous study. In the reaction column, iron scraps (1000 g) and GAC (500 g) were combined and inserted together. Thus, many microscopic galvanic cells were formed between the iron and carbon, when these particles were in contact with wastewater. The wastewater flowed into the column from the bottom, and air at a flow rate of 0.3 L/min was aerated into the bottom of the column through an air pump, that was important to maintain the iron scraps and GAC in the reaction column was consistently diversed and raised the removal efficiency of organic contaminants. Mainly, the pH solution and reaction time enhanced the treatment efficiency of internal micro-electrolysis on top of the aeration.

Based on the result from the research, the rate on TN removal in HMBR was 92.9%, larger than the control CMBR. However, in both MBRs, the results of COD and NH_3–N removal rates were superior and stable with a percentage larger than 90%. The iron had useful impacts on the structure and property of the flocs. With the increasing particle size, the flocs' settling ability and compatability also become better, thus prohibiting the formation of membrane fouling in HMBR. However, in this IE/MBR system,

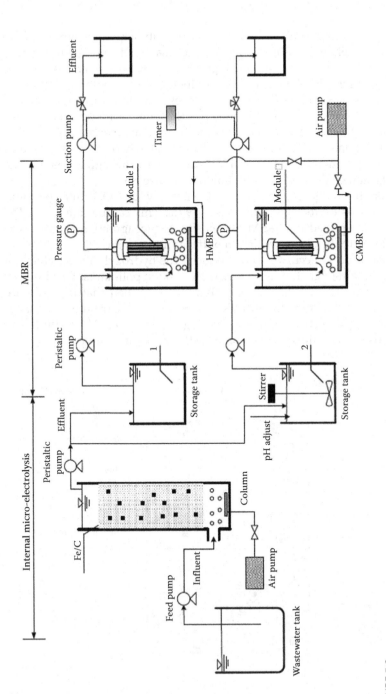

FIGURE 5.2
Schematic diagram of experimental process for enhanced MBR by internal micro-electrolysis. (Adapted from Qin, L. et al., 2012. *Chemical Engineering Journal*, 210, 575–584.)

further investigation was still needed in controlling inorganic membrane fouling.

Another advanced MBR technology is the submerged membrane electro-bioreactor (SMEBR) which applied electrical field and microfiltration in a nutrient-removing activated sludge process for wastewater treatment. The combination of biological, electrokinetic, and membrane filtration processes in one hybrid SMEBR was first investigated by Elektorowicz et al. (2009). In the SMEBR, wastewater entered the biological treatment zone I (i.e., between the reactor wall and the outer electrode), then passed through the electrical zone II (i.e., between the electrodes), where it was exposed to electrokinetic phenomena and was finally filtered out through the membrane module (Hasan et al., 2014). This is shown in Figure 5.3.

Based on the research, the removal efficiency of COD, nitrogen, and phosphorus was quite high, around 92%–99%. This can be due to the DC electrical field applied in the SMEBR; *in situ* Al^{3+} (coagulation agents) were generated due to the electrooxidation of the aluminum anode.

Oxidation process occurred and produced the hydroxyl radical which, as a strong oxidizing agent, might react with organic pollutants. Besides that, Ibeid et al. (2010) and Hasan et al. (2012b) also claimed that the presence of electrokinetics within the SMEBR has positively contributed to changing the sludge properties and thus lowering fouling rates. The SMEBR also showed promising ability to overcome costly frequent membrane cleaning.

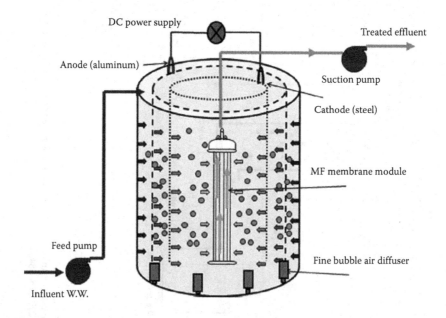

FIGURE 5.3
The SEMBR. (Adapted from Elektorowicz, M., Bani-Melhem, K., and Oleszkiewicz, J . 2009. Submerged Membrane Electro-Bioreactor—SMEBR, US Patent 12553, 680.)

Additionally, the generated sludge from the SMEBR, including the inorganic biosolids have no toxic elements (Hasan et al., 2012a) suggesting that the sludge production volume in the SMEBR was minimized, and therefore reducing the sludge handling and management costs accordingly. Moreover, since the SMEBR can be built as a modular system, the authors believe that this technology might be used in different sizes of plants including individual homes.

5.4 Growth of MBR Market

The global market for MBR technology is forecast to extend to \$888 million by the year 2017 due to rigid effluent discharge standards, with an emphasis on water reuse and recycling for freshwater preservation (Global MBR market, 2012). Besides technological growth, the necessity to change traditional equipment and increasing application across several end-use markets are also factors to nourish market growth. Global industry analysis claimed that the use of MBRs is fast enhancing in the world in terms of research, technology, and commercial applications. Due to several applications and the legislative burden, the market is distinguished by unique growth rates for various markets. For some time, the benefits offered by MBR technology have been popular. In retaining the biomass, MBR consists of a CAS process combined with membrane separation. According to dissolved constituents like organic matter and ammonia, both are importantly reduced by biotreatment, hence MBRs are capable of producing treated waters of higher purity (Santos et al., 2011).

The global MBR market tends to increase at a comparatively faster rate compared to other kinds of membrane systems used in wastewater treatment, with MBR technology discovering growing use in municipal, marine, and industrial applications. The requirement to replace older infrastructure for water treatment and the installation of new infrastructure is targeted at enhancing growth in the future. Based on a BCC report (BCC, 2008), the market entrance of MBR technology is reported as increasing by an average of 11.6%–12.7% per annum since the turn of the millennium (Srinivasan, 2007). According to the BCC study, this market is increasing faster than the larger market for advanced wastewater treatment equipment, at about a 5.5% annual growth rate, and more quickly than the market for other types of membrane systems, which are increasing at rates from 8% to 10%, relying on technology. By organizing local supply networks, expanding local partners, and also training and using a local workforce, MBR companies might accomplish cost-efficiencies. New ground is also hoped to be covered by significantly promoting their products' small footprint and high-efficiency treatment in processes like nutrient removal.

In renovating old plants, the small footprint of recent technologies is useful. The latest MBRs are also qualified in purifying water as per the rigid requirements in terms of water reuse and nutrient removal. MBRs are then targeted to discover growing use in industrial applications over the coming years, especially for water reuse, predicts the study. With decreasing capital costs providing additional force, the rebirth of economic growth is targeted to result in the extensive adoption of the technology. The municipal sector shows the widest end-use employment for MBRs.

MBR contributes to useful savings over operational expenditures due to current MBR system designs that are energy efficient. Compared to other technologies they are also relatively cheap. MBR systems have a competitive edge over conventional water purification systems like reverse osmosis because of these and other technological characteristics. Moreover, MBR systems do not need chemicals for the filtration function. For the establishment of water treatment plants, limited space acts as the main challenge to fulfill the requirement of all communities. In the global MBR market place, packaged MBR technology is one of the rapidly growing sections. Potential growth opportunities were offered to small vendors and new market entrants. In the United States and Europe, it is a fast emerging segment with several well organized market players.

5.5 Summary

Since individuals, communities, industries, nations, and their national institutions struggle for ways to maintain this essential resource ready and suitable for use, water treatment has been a field of global concern. Over the past 10 years, MBRs have grown as an efficient way of converting various types of wastewater into high-quality effluent useful for discharge into the environment, and rising into a reusable product. Besides that, significantly high-quality product water is useful for discharge into the environment to a publicly owned treatment plant or for recycling within the industrial plant's processes in industrial applications, where rigid discharge standards are in force, also offered by MBR. In-plant recycling can be useful for industries struggling with municipal, agricultural or other, more essential industrial users to purchase fresh water. These various end-users may also strive for effluent discharge permits. A decentralized small-scale wastewater treatment for remote or isolated communities, campsites, tourist hotels, or industries not connected to municipal treatment plants was also provided by MBR. MBR technology would supply a decentralized, strong, and cost-effective treatment for managing high-quality effluent for developing nations with vast areas with no sewerage systems. Excellent retrofit capability was also suggested by MBR for increasing or upgrading existing conventional wastewater treatment plants.

References

Achilli, A., Marchand, E. A., and Childress, A. E. 2011. A performance evaluation of three membrane bioreactor systems: Aerobic, anaerobic, and attached-growth. *Water Science and Technology*, 63(12), 2999–3005.

BCC. 2008. *Membrane Bioreactors: Global Markets*, BCC Report MST047B.

Cheng, H., Xu, W., Liu, J., Wang, H., He, Y., and Chen, G. 2007. Pretreatment of wastewater from triazine manufacturing by coagulation, electrolysis, and internal microelectrolysis. *Journal of Hazardous Materials*, 146, 385–392.

Elektorowicz, M., Bani-Melhem, K., and Oleszkiewicz, J. 2009. *Submerged Membrane Electro-Bioreactor—SMEBR*, US Patent 12553,680.

Friha, I., Karray, F., Feki, F., Jlaiel, L., and Sayadi, S. 2014. Treatment of cosmetic industry wastewater by submerged membrane bioreactor with consideration of microbial community dynamics. *International Biodeterioration & Biodegradation*, 88, 125–133.

Global MBR market forecast to reach $888 million by 2017. 2012. *Membrane Technology*, 2012(1), 8.

Hasan, S., Elektorowicz, M., and Oleszkiewicz, J. 2012a. Pilot submerged membrane electro-bioreactor (SMEBR) for COD, nutrients and heavy metals removal. In: *10th International Scientific and Technical Conference "Water Supply and Water Quality,"* Stare Jablonki, Poland, September 9–12, 2012.

Hasan, S. W., Elektorowicz, M., and Oleszkiewicz, J. 2012b. Correlations between transmembrane pressure (TMP) and sludge properties in submerged membrane electro-bioreactor (SMEBR) and conventional membrane bioreactor (MBR). *Bioresource Technology*, 120, 199–205.

Hasan, S. W., Elektorowicz, M., and Oleszkiewicz, J. A. 2014. Start-up period investigation of pilot-scale submerged membrane electro-bioreactor (SMEBR) treating raw municipal wastewater. *Chemosphere*, 97, 71–77.

Hoinkis, J., Deowan, S. A., Panten, V., Figoli, A., Huang, R. R., and Drioli, E. 2012. Membrane bioreactor (MBR) technology—A promising approach for industrial water reuse. *Procedia Engineering*, 33, 234–241.

Howell, J. A. 2004. Future of membranes and membrane reactors in green technologies and for water reuse. *Desalination*, 162, 1–11.

Huang, X., Wei, C.-H., and Yu, K.-C. 2008. Mechanism of membrane fouling control by suspended carriers in a submerged membrane bioreactor. *Journal of Membrane Science*, 309, 7–16.

Ibeid, S., Elektorowicz, M., and Oleszkiewicz, J. 2010. Modification of activated sludge characteristics due to applying direct current (DC) field. In: *Proceedings IWA Water World Congress*, Montreal, Canada, September 20–27, 2010.

Jamal Khan, S., Zohaib Ur, R., Visvanathan, C., and Jegatheesan, V. 2012. Influence of biofilm carriers on membrane fouling propensity in moving biofilm membrane bioreactor. *Bioresource Technology*, 113, 161–164.

Katayon, S., Megat Mohd Noor, M. J., Ahmad, J., Abdul Ghani, L. A., Nagaoka, H., and Aya, H. 2004. Effects of mixed liquor suspended solid concentrations on membrane bioreactor efficiency for treatment of food industry wastewater. *Desalination*, 167, 153–158.

Liu, J., Liu, L., Gao, B., and Yang, F. 2012. Cathode membrane fouling reduction and sludge property in membrane bioreactor integrating electrocoagulation and electrostatic repulsion. *Separation and Purification Technology*, 100, 44–50.

Liu, J., Liu, L., Gao, B., and Yang, F. 2013. Integration of bio-electrochemical cell in membrane bioreactor for membrane cathode fouling reduction through electricity generation. *Journal of Membrane Science*, 430, 196–202.

Melin, T., Jefferson, B., Bixio, D., Thoeye, C., De Wilde, W., De Koning, J. et al., 2006. Membrane bioreactor technology for wastewater treatment and reuse. *Desalination*, 187, 271–282.

Qin, L., Zhang, G., Meng, Q., Xu, L., and Lv, B. 2012. Enhanced MBR by internal micro-electrolysis for degradation of anthraquinone dye wastewater. *Chemical Engineering Journal*, 210, 575–584.

Radjenovic, J., Matosic, M., Mijatovic, I., and Petrovic, M. 2008. Membrane bioreactor (MBR) as an advanced wastewater treatment technology. Emerging contaminants from industrial and municipal waste. *The Handbook of Environmental Chemistry*. Springer Berlin Heidelberg, 37–101.

Rosenberger, S., Krüger, U., Witzig, R., Manz, W., Szewzyk, U., and Kraume, M. 2002. Performance of a bioreactor with submerged membranes for aerobic treatment of municipal waste water. *Water Research*, 36, 413–420.

Sadri, S., Cicek, N., and Van Gulck, J. 2008. Aerobic treatment of landfill leachate using a submerged membrane bioreactor—prospects for on-site use. *Environmental Technology*, 29, 899–907.

Santos, A., Ma, W. and Judd, S. J. 2011. Membrane bioreactors: Two decades of research and implementation. *Desalination*, 273, 148–154.

Sombatsompop, K., Visvanathan, C., and Ben Aim, R. 2006. Evaluation of biofouling phenomenon in suspended and attached growth membrane bioreactor systems. *Desalination*, 201, 138–149.

Srinivasan, J. 2007. MBR still growing in EU wastewater treatment market. *Water and Waste International*, 22, 43–44.

Tchobanoglous, G., Darby, J., Ruppe, L., and Leverenz, H. 2004. Decentralized wastewater management: Challenges and opportunities for the twenty-first century. *Water Science and Technology: Water Supply*, 4(1), 95–102.

Tiranuntakul, M. 2012. Membrane application as an advanced wastewater treatment. *Ladkrabang Engineering Journal*, 29(3), 7–12.

Yang, W., Cicek, N., and Ilg, J. 2006. State-of-the-art of membrane bioreactors: Worldwide research and commercial applications in North America. *Journal of Membrane Science*, 270, 201–211.

Yigit, N. O., Uzal, N., Koseoglu, H., Harman, I., Yukseler, H., Yetis, U. et al., 2009. Treatment of a denim producing textile industry wastewater using pilot-scale membrane bioreactor. *Desalination*, 240, 143–150.

6

Integration of Membrane Bioreactor with Various Wastewater Treatment Systems

Chin Hong Neoh, Zainura Zainon Noor, Cindy Lee Ik Sing, Florianna Lendai Michael Mulok, and Noor Salehan Mohammad Sabli

CONTENTS

6.1 Introduction

The increasing concerns regarding water resources are becoming an important issue as water stress due to climate change, pollution, and urbanization have been seen throughout the world. Though membrane bioreactor (MBR) has made enormous progress and has become a promising approach in wastewater treatment, there are still several limitations to conventional MBR and thus integrated MBR technology for wastewater treatment has emerged. MBR alone is not sufficient to guarantee an adequate reduction of pollutants from wastewater. This chapter aims to provide a consolidated review on the current state of research for the integrated MBR system with other technologies for wastewater treatment and help to sustain the treatment process itself. It is expected to provide future prospects to implement integrated MBR as a promising wastewater treatment technology. The valorization of wastewater in integrated MBR would reduce pollution and more importantly open a new source of energy.

Remarkable progress has been achieved in the application of the MBR process to wastewater treatment and reclamation. Rapid population expansion in developing countries has caused existing conventional wastewater treatment plants to become overloaded and there will be no space available for their expansion (Haandel and Lubbe, 2011). In view of this, MBR has attracted growing interest as it has some distinct advantages of smaller footprint, less sludge production, higher separation efficiency, and highly improved effluent quality as compared to conventional activated sludge treatment (Mutamim et al., 2012; Tan et al., 2015). MBR is also able to retain small molecular weight organic micropollutants compared to the conventional activated sludge process. MBR with ultrafiltration membrane is also able to retain some types of viruses (Rodríguez et al., 2011).

Legislation, increasing level of water stress, and growing confidence in the performance of MBR are the key market drivers which contribute to the expansion of MBR market. According to the Market Research Report, the Asia-Pacific market share in the global MBR revenue market is 38%, followed by Europe (17%). The global MBR market is rising at a compound annual growth rate of 22.4% and is expected to reach $3.44 billion by 2018 (Royan, 2016). The MBR market is growing fast and has contributed to the larger market for wastewater treatment equipment including physical, chemical, and biological treatments, and membrane filtration. Meanwhile, MBR is also the largest market for other membrane systems treating wastewater such as microfiltration (MF), ultrafiltration, nanofiltration, and reverse osmosis (RO) (Hanft, March 2011; SBI, October 05, 2012). Even though some wastewater treatment systems may be capable of treating industrial wastewater to meet current disposal requirements and producing water for basic uses in the industry, the treated effluent would need to be further polished by using integrated MBR for applications that need high grade water.

Extensive review papers are available for MBR in wastewater treatment technology. Ylitervo et al. (2013) studied the potential of MBR that is not focused on wastewater treatment technology, but on ethanol and biogas production in fermentation technology. Besides, applications and limitations of MBR in the treatment of high strength industrial wastewater have also been studied (Mutamim et al., 2012, 2013). Wang et al. (2014) made a critical review on physical, chemical, and biological cleaning in MBR and proposed the procedures for determining proper cleaning protocols. Goh et al. (2014) studied the potential applications and possible configurations for membrane distillation bioreactor (MDBR) while Wang and Chung (2015) studied the development, configuration design, and application of MDBR. In contrast, Tijing et al. (2015) studied fouling and its control in MDBR. Pellegrin et al. (2013) studied the membrane processes for municipal and industrial applications which covers the pretreatment, MBR configuration, membrane fouling, fixed film and anaerobic MBR, membrane technology advances, and modeling. Most review articles related to integrated MBR are on anaerobic MBR (AnMBR) which has the advantage of reducing organic matter and producing energy under anaerobic processes. For example, Ozgun et al. (2013) studied the integration, limitations, and expectations of AnMBR for municipal wastewater treatment. Another study on AnMBR was conducted by Phattaranawik et al. (2009) with special emphasis on performance and bottlenecks in terms of its application in industrial scale. Skouteris et al. (2012) studied the performance of AnMBR which focuses on the comparison with other wastewater treatment technologies, energy recovery, and membrane fouling issues. There are some upon other studies conducted various types of integrated MBR wastewater treatment and valorization. This chapter aims to provide a review of the current research for the integrated MBR system for wastewater reclamation. It is the first study ever in the integration of MBR with other treatment technologies such as advanced oxidation processes (AOPs), granulation technology, reverse and forward osmosis (FO), MDBR, and hybrid moving bed biofilm reactor-membrane bioreactor (hybrid MBBR-MBR) to enhance the treatment efficiency of wastewater. The other important parts of this chapter are focused on the sustainable development of MBR for the production of biofuel, generation of electricity, and recovery of nutrients for income generation and also contribute to the environment.

6.2 Advanced Oxidation Processes and Electrocoagulation Processes

AOPs in general are well known for their capacity to remove many organic contaminants. AOPs are able to convert recalcitrant pollutants into biodegradable intermediates that can be degraded in a biological process. However, it is negatively affected by suspended solids that act as scavengers toward hydroxyl radicals which are formed by ozone decomposition in water. In this view, MBR offers unique opportunities in terms of suspended solid-free effluent and thus enhances the efficiency of ozonation. AOPs can be used as pre- or postbiological treatment of wastewater. Pretreatments have been proven useful in the case of wastewater which contains small amounts of biodegradable organics and a large amount of recalcitrant compounds. In contrast, posttreatment results are preferable when biodegradable compounds are greater than those of recalcitrant compounds (Laera et al., 2011).

Mascolo et al. (2010) studied the effective organic degradation from pharmaceutical wastewater by an integrated process of MBR and ozonation. The reactor was set up by placing the ozonation reactor in the recirculation stream of the MBR effluent. The organic compound (acyclovir) in the effluent was removed up to 99% from the MBR step and ozonation allowed to further remove 99% of the MBR effluent. For several organics identified in the wastewater, the efficiency of the MBR treatment improved from 20% to 60% when the ozonation was placed in the recirculation system. This study found out that MBR-ozonation system gave results comparable to those obtained by the two separate treatment systems. In contrast, López et al. (2010) studied the integration of solar photo catalysis followed by MBR for pesticide degradation. The permeate obtained in the coupled system was ready-to-use high quality water, with the absence of pesticides and solids, and turbidity values (nephelometric turbidity units [NTU]) below 0.5. The results demonstrated that this system is able to treat the pesticide mixture without adding a carbon source. Laera et al. (2011) investigated the integration of MBR with either ozonation or UV/H_2O_2 process by placing chemical oxidation in the recirculation stream of the MBR. The study reported the removal of synthetic wastewater containing nalidixic acid which was used in treating urinary tract infections. MBR alone is not efficient in removing the degradation products of the nalidixic acid as those compounds would pass through the MBR. However, integration of ozonation completely removed the degradation products in the step of chemical oxidation. An integrated thermophilic submerged aerobic membrane bioreactor (TSAMBR) and electrochemical oxidation technology using a Ti/SnO_2–Sb_2O_5–IrO_2 electrode was developed for treatment of pulp and paper effluent (Qu et al., 2012). The integrated oxidation processes completely decolorized the effluent and further enhanced the removal of chemical oxygen demand (COD). High-quality effluent can be produced through the integrated process and has the potential for the direct reuse in the mill. Giacobbo et al. (2014) investigated the integrated MBR-photoelectrooxidation (MBR–PEO) for tannery wastewater treatment. The MBR is responsible for the remaining biochemical oxygen demand (BOD) removal, while the refractory matter (contributed to COD) is removed by PEO. The treated wastewater could be recycled for the tanning and retanning steps. Fouling does not penetrate in the membrane pores and the membrane performs well after 360 h of work without membrane cleaning. Lamsal (2012) examined the fouling in nanofiltration membranes and explored various pretreatment strategies to reduce the fouling process. The author found out that AOP pretreatments with NF membrane resulted in an improved permeate flux but not permeate quality of the NF membrane. Merayo et al. (2013) stated that higher COD removal was achieved treating the pulp mill effluent from the recycled MBR followed by AOPs. The remaining recalcitrant compounds could be more efficiently ozonized when the effluent was treated with MBR.

Electrocoagulation as can be seen in Figure 6.1 is a technique that generates the coagulants *in situ* by dissolving electrically through either aluminum or iron ions from electrodes. The metal ions are generated at the anode while hydrogen gas is generated at the cathode. The released hydrogen gas will assist in floating the flocculated particles to the surface (Torres-Sánchez et al., 2014). Bani-Melhem and Smith (2012) compared the performance of electrocoagulation-MBR and MBR alone for the treatment of grey water. The results showed that electrocoagulation-MBR achieved up to 13% reduction of membrane fouling at a higher reduction of COD, turbidity, phosphate, and color compared to MBR alone. However, ammonia nitrogen was removed more effectively by MBR alone and this might be due to sensitivity of nitrifying bacteria by the aluminum concentration produced. Keerthi et al. (2013) investigated three different combinations of treatment technology (electrocoagulation combined with MF, MBR, electrocoagulation integrated with MBR). The electrocoagulation integrated with MBR showed the best performance

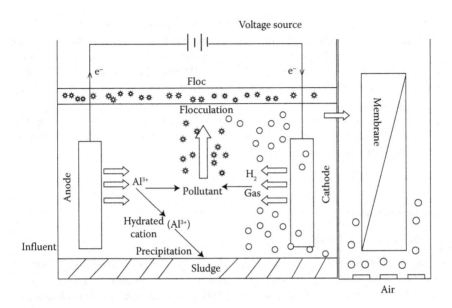

FIGURE 6.1
Schematic diagram of the integrated electrocoagulation-MBR with an external side-stream membrane.

in fouling reduction and the removal of COD and metal in tannery wastewater. This kind of integrated MBR is suitable for the treatment of wastewater similar to tannery wastewater aiming at zero discharge. Another study by Vijayakumar et al. (2014) reported that integrating electrocoagulation using aluminum as anode and stainless steel as cathode with MBR is very effective in the removal of metal compared to conventional MBR (without any integrated system). The integrated MBR was found to be efficient in removal of Cr, Cu, and Zn with an average removal efficiency of more than 90% (initial concentration was 25 mg/L). This system also enhanced permeation flux and improved membrane life. Alizadeh Fard et al. (2013) focused on the application of electrocoagulation, electrofenton, and electron–Fenton processes for excess sludge treatment in MBR. Technical and economical investigations conducted in this study have shown that electrofenton–Fenton system was the best one. In the control of MBR, the constant sludge yield was approximately 0.1 g when mixed with suspended solid liquor (MLSS)/g of COD, however, when MBR is combined with the sludge oxidation process, the excess sludge production was almost zero. The oxidation process was found to effectively reduce membrane fouling and prolong the membrane chemical cleaning cycle. Still, the effect of sludge oxidation and total phosphorus removal was insignificant for both control and MBR–electrofenton processes.

To sum up, integrating MBR with AOPs and electrocoagulation were found to be effective in the mitigation of membrane fouling, removal of recalcitrant compounds, as well as colored compounds and metal. In other words, this system was effective in the treatment of pharmaceuticals wastewater, and colored wastewater such as textile wastewater. Nevertheless, the excess sludge production, cost and optimization of parameters such as sludge retention time (SRT) and hydraulic retention time (HRT) needs to be further explored and investigated. APOs and electrocoagulation were found to be effective for wastewater treatment and reuse as they are able to produce a high quality of effluent. However, more emphasis needs to be placed on investigating the effect of AOPs and electrocoagulation in detecting pathogens in the effluent, and feasibility thereof.

6.3 Anaerobic MBR

Integrated MBR with anaerobic digestion is considered the "zero" concept for MBR, targeting the effective treatment of wastewater and recovery of energy and materials (Solomou et al., 2014). Biogas is renewable fuel and can be used as an alternative to fossil fuels. The production of biogas depends on the activity of methanogen. MBR for biofuel production is still a new concept whereby only a few industries employ anaerobic biological process for wastewater treatment (Ylitervo et al., 2013). Methanogen has a very slow growth rate and is easier to wash out from the conventional anaerobic treatment system. The anaerobic process is complex compared to aerobic processes. There are high chances of process failure due to the presence of inhibitory substances, such as heavy metals, chlorinated hydrocarbons, and cyanides that are present in feeding wastewater or sludge. The net biomass production must exceed the net biomass loss for biological treatments to function properly. However, in conventional anaerobic biological treatments, the net biomass lost to the effluent is higher due to the poor settling characteristics of the biomass. Besides that, conventional anaerobic biological treatments were also not effective in removing residual levels of soluble and colloidal contaminant with poorer gas capture, poorer odor control, limited ability to capture nutrients, and expensive desludging processes (Gallucci et al., 2011; Jensen, 2015). In view of this, AnMBR has gained attention as it has potential for energy recovery and retains the biomass for efficient treatment. The integration of membrane also results in a good quality effluent that has significantly lower amount of particles and pathogens (Ozgun et al., 2013). Table 6.1 shows the types of AnMBR and membrane yield generated in each AnMBR.

Sheldon et al. (2012) studied the utilization of an anaerobic expanded granular sludge bed (EGSB) for the treatment of paper mill effluent followed by a posttreatment side-stream MBR (ultrafiltration membrane). The EGSB system successfully removed 65%–85% of COD and satisfactory improved results were obtained with an overall MBR reduction in COD

TABLE 6.1

Types of AnMBR and Membrane Yield Generated in Each AnMBR

Authors	Types of AnMBR	Wastewater	Methane Yield	Membrane
Gao et al. (2014)	Integrated anaerobic fluidized-bed MBR	Domestic wastewater	180 L methane per kg $COD_{REMOVED}$	Hollow-fiber membrane
Solomou et al. (2014)	MBR-anaerobic digestion system	Municipal wastewater	456.7 L/kg volatile solid	Polyethersulfonate membrane
Wei et al. (2014)	Mesophilic AnMBR	Synthetic municipal wastewater	300 mL/gCOD	Hollow-fiber PVDF UF membrane
Youngsukkasem et al. (2014)	Reverse MBR	Synthetic organic medium	0.9 mmol	PVDF membrane
Chen et al. (2014)	FO submerged AnMBR	Synthetic wastewater	0.21 L/gCOD	Flat-sheet cellulose triacetate (CTA) membranes
Sheldon et al. (2012)	Anaerobic/aerobic hybrid side-stream MBR	Paper mill effluent	28.6 L/day	ultrafiltration membrane
Jensen (2015)	AnMBR	Red meat processing wastewater	360L methane per kg COD	Submerged hollow-fiber membrane

of 96%. An upward trend of methane production was noticed when the HRT decreased (from 10.3 to 7.7 h) and the organic loading rate (OLR) increased (from 4.8 to 5.5 kg COD). Integrated anaerobic fluidized-bed membrane bioreactor (IAFMBR) with granular activated carbon (GAC) was developed to treat domestic wastewater with energy recovery. At HRT for 6 h, methane yield was 180 L; and conversion of COD into methane in biogas was 53%. The reactor contained two parts; the outer part performs as anaerobic fluidized-bed reactor with GAC as carriers and the inner part serves as AnMBR (Gao et al., 2014). An integrated system for the treatment of wastewater and biodegradable organic waste was examined (Solomou et al., 2014). Under optimized conditions, this system was able to remove 99% of ammoniacal nitrogen, 95% of nitrogen, and 96% of COD. Energy balance showed that the system required 5% of heat and 3.5% of electricity from the total energy of the produced biogas to maintain its operation. This resulted in net electricity production of 31.5% and the excess heat was 50% after considering the heat and energy requirements of the anaerobic digestion unit. Wei et al. (2014) studied the comprehensive effects of OLRs on methane production as well as on organic removal and biomass production to optimize AnMBR operation. A very steady and high COD removal was achieved over a broad range of volumetric OLR. The AnMBR achieved high methane production of over 300 mL/g COD at a high sludge OLR of over 0.6 g COD/gMLVSS/day. The authors concluded that the integration of the heat pump and FO into the mesophilic AnMBR process would be a promising way to net energy recovery from typical municipal wastewater, especially in a temperate area. Youngsukkasem et al. (2014) studied the rapid bio-methanation of syngas in a reverse membrane bioreactor (RMBR) by using membrane-encased microorganisms. Recalcitrant compounds such as lignin and plastic wastes cannot be decomposed by the microorganisms in a conventional anaerobic process. However, thermochemical processes have the potential to convert this kind of waste and nondegradable residue into intermediate gas, called syngas. The digesting sludge encased in the polyvinylidene difluoride (PVDF) membranes was able to convert syngas into methane and displayed a similar performance as the free cells in batch fermentation. At thermophilic conditions (55°C), there was a higher conversion of pure syngas and codigestion using the encased cells compared to mesophilic conditions. A submerged AnMBR with FO membrane was studied by Chen et al. (2014). The AnMBR was operated at 25°C using synthetic wastewater as substrate. Due to the membrane fouling and the increasing salinity in the AnMBR, the water flux was reduced. However, the authors claimed that the level of salinity in the reactor was not a concern in terms of inhibition or toxic effects and an average value of 0.21 L CH_4/g COD was obtained. The removal rates for ammoniacal nitrogen and total phosphorus were up to 60% and 100%, respectively, exhibiting the excellent interception of FO membrane. Jensen (2015) studied the treatment of wastewater and nutrient recovery from red meat processing facilities using an AnMBR pilot plant. The AnMBR removed over 90% of COD consistently and produced approximately 360 L of methane per kg COD. Economic comparisons show that the payback of AnMBR is comparable to a conventional anaerobic lagoon due to increased gas capture resulting in improved energy recovery.

Jensen (2015) estimated that in an AnMBR, 10% of the capital costs are in membranes and 50% in the vessel. Biogas produced through AnMBR can lower the cost further by using the gas to feed a boiler, and/or a trigeneration system to produce heat, power, and cooling. Pretel et al. (2014) investigated the operating cost of an AnMBR treating sulfate-rich wastewater. Operating at high ambient temperature and/or high SRT allows significant energy savings (minimum energy demand: 0.07 kW h/m³). While low/moderate sludge productions were obtained (minimum value: 0.16 kg TSS/kg $COD_{REMOVED}$), which further enhanced the operating cost (minimum value: €0.01 per m³). Each cubic meter of biogas contains the equivalent of 6 kW h or 21.6 MJ of energy and approximately 35% of the total energy is converted to electricity while the rest is converted into heat, some of which can be recovered for heating applications (McCabe et al., 2014).

The challenge in biogas recovery is the limited amount of biogas produced in small-scale treatment plants. Most of the methane produced in the treatment plant is lost through the effluent of the plant. The methane concentration of about 16 mg/L (equivalent COD 64 mg/L) is expected in the effluent caused by the partially high pressure of methane gas inside the treatment plant (Adrianus and Gatze, 1994). This issue must be addressed for ideal application of AnMBR. Besides, continuous process optimization should be conducted as the payback period of an AnMBR remains comparatively high (Jensen, 2015). Last but not least, the competition between methanogen and sulfate in reducing bacteria for the available substrate is another issue faced in the AnMBR. The available COD for methanization would reduce when there is significant sulfate content in the effluent (Pretel et al., 2014).

6.4 Microbial Fuel Cell

A microbial fuel cell (MFC) is a bioelectrochemical device which uses microorganisms as catalyst to convert the chemical energy in organic matters into electrical energy. It is generally composed of two chambers; an anode chamber where the oxidation of organic compounds takes place under anaerobic condition and a cathode chamber where the oxygen or ferricyanide is reduced under aerobic condition (Figure 6.2) (Min and Angelidaki, 2008). The two anode and cathode chambers are normally separated by a permeable

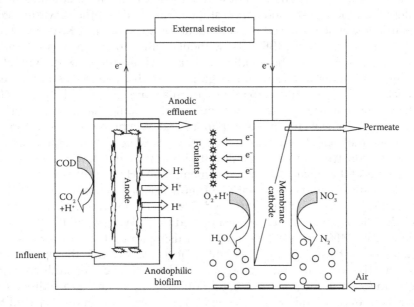

FIGURE 6.2
Schematic diagram of MFC–MBR and mechanism of electric field on membrane fouling mitigation. Electrons generated in anode are transferred to membrane cathode and exerts additional repulsion force to membrane foulants.

membrane. In other words, an MFC acts as an inexpensive biosensor and can provide clean and safe energy, quiet performance, low emissions, and ease in operating, apart from treatment of wastewater (J. Wang et al., 2014). One of the restrictions in the application of MBR is the high energy consumption, estimated at 0.8–1.1 kWh/m^3 (J. Li et al., 2014). Thus, the integration of MFC with MBR is recommended since the energy consumption of MBR can be further lowered. However, MFC alone leads to low efficiency treatment and poor effluent quality due to limited biomass retention (Logan, 2008). Effluent from MFC alone still contains a certain amount of suspended solids and the remaining contaminants need further treatment before being discharged. The combination of the MBR–MFC system, known as electrochemical membrane bioreactor (EMBR) or membrane bioelectrochemical reactor (MBER) or bioelectrochemically assisted membrane bioreactor (BEAMBR) offers a convincing option for wastewater treatment and energy recovery. The types of cathode, anode, membrane, and power density generated in each MFC–MBR system were listed in Table 6.2.

Z. Wang et al. (2013) conducted a study related to MBR–MFC and stated that only 28% of COD was removed in the anode chamber while the majority of organic pollutants (total removal of COD was 85%) were removed in cathode chamber. Y.-P. Wang et al. (2012) claimed that they developed a more practical MBR–MFC integrated process in which they were trying to reduce the investment and operating cost. The authors used the aeration tank of the MBR as a cathode chamber, low-cost nylon mesh was adopted as filter material, and low-cost anode and cathode was used. Synthetic wastewater containing acetate was used as an influent for the system. Y.-K. Wang et al. (2013) studied the use of MBR–MFC without aeration for energy recovery and wastewater treatment. The anodic chamber was inoculated with synthetic wastewater containing acetate. The graphite cloth cathode was immerged in a laboratory scale MBR for 5 min to inoculate nitrifiers and denitrifiers to enhance the operation process. Effluent from the anodic chamber penetrated through the nonwoven cloth (as separator and membrane) and air cathode, and was finally discharged from the system. Despite the air cathode being easier for construction and significantly improving the power density, the drawbacks of the air cathode included membrane pH gradient, water leaking through cathode, and accumulation of inorganic salt deposits. The authors claimed that those problems improved through the unique configuration of their EMBR. Another study was conducted by Z. Wang et al. (2013) where a new MBR–MFC process was developed using an up-flow reactor design. The lower part was the anode chamber of the MFC while the upper part was an MBR installed with a stainless-steel membrane. The authors used the stainless-steel membrane as solid liquid separator and as a cathode for the MFC. Graphite rods were fastened on the membrane and connected to the external resistor by copper wire. The authors claimed that this type of MBR–MFC is easier for operation and maintenance. Again, the synthetic wastewater containing acetate as influent was used by the authors. Ge et al.

TABLE 6.2

Types of Cathode, Anode, Membrane, and Power Density Generated in Each MFC–MBR System

Membrane	Anode	Cathode	Maximum Power Density (W/m³)	Authors
PVDF membrane	Chamber was made of nonconductive plexiglass plate	Carbon brushes	2.18	Tian et al. (2015)
Stainless-steel mesh	Graphite rod	Stainless-steel mesh	4.35	Y.-K. Wang et al. (2011)
Nylon mesh with pore size of 74 μm	Activated carbon fiber	Carbon felt	6	Y.-P. Wang et al. (2012)
Nonwoven cloth	Graphite felt	Graphite felt	7.6	Y.-K. Wang et al. (2013)
Flat-sheet membrane made of stainless-steel mesh	Graphite felt	Graphite rod and stainless-steel membrane	8.62	Z. Wang et al. (2013)
Stainless-steel mesh	Graphite rod	Stainless-steel mesh	1.43	Y.-K. Wang et al. (2013)
PVDF hollow-fiber membrane	Carbon brush	Carbon cloth coated with 10% platinum (Pt)	2 (wastewater)	Ge et al. (2013)
PVDF hollow-fiber membrane	Carbon cloth	Carbon cloth coated with 10% Pt	0.15	J. Li et al. (2014)
Flat-sheet membrane made of stainless-steel mesh	Graphite rod	Flat-sheet membrane made of stainless-steel mesh	8.6	Huang et al. (2014)
Stainless-steel membrane	Graphite granules	Stainless-steel mesh		Liu et al. (2014)
Polyester filter cloth, modified by *in situ* formed PANi-PA	Graphite granules	Polyester filter cloth, modified by *in situ* formed PANi-PA	0.78	N. Li et al. (2014)
Hollow-fiber MF	Carbon cloth	Contained 10% platinum and poly tetra fluoro ethylene (PTFE) diffusion layers	51 mW/m²	Su et al. (2013)

(2013) investigated the use of hollow-fiber membrane bioeletrochemical reactor for domestic wastewater treatment. The MBR–MFC was constructed with a tube made of cation exchange membrane (CEM). The CEM tube formed an anode compartment and the cathode was directly exposed in the air to avoid active aeration. The MBR–MFC was first operated with synthetic wastewater containing acetate, producing maximum electricity at 0.038 kWh/m^3. The MBR–MFC proceeded using domestic wastewater and produced the maximum electricity at 0.025 kWh/m^3. The wastewater gave lower electricity as the COD is 1/3 lower than the synthetic wastewater. Although this kind of MFC–MBR gave a lower energy consumption (no aeration required), this kind of reactor has a limited nitrogen removal and requires postprocesses to remove or recover nutrients like other anaerobic treatments (Smith et al., 2012). J. Li et al. (2014) studied the advancing MBR–MFC with hollow-fiber membranes installed in the cathode compartment. The cathode compartment was aerated with air. The energy production in the MBR–MFC was 0.011–0.039 kWh/m^3 from the synthetic wastewater containing acetate and 0.032–0.064 kWh/m^3 from the cheese whey wastewater. Compared to MBR and AnMBR fed with the cheese whey wastewater, MBR–MFC used by J. Li et al. (2014) achieved a better removal of COD and total nitrogen (TN). Huang et al. (2014) studied two MBR–MFCs operated under closed-circuit and open-circuit modes. Maximum power density of the MBR under closed-circuit operation was as high as 8.6 W/m^3, indicating that the biochemical MBR did not only achieve efficient wastewater treatment but also power production. To the best of our knowledge, this is the only study on identification of microbial communities on MBR–MFC by high-throughput 454 pyrosequencing. Under closed-circuit condition, *Bacteroidetes* was the dominant phyla in the anode samples while *Protebacteria* was the most dominant phyla in the cathode samples. N. Li et al. (2014) studied a simple and low-cost MBR–MFC system without using any noble metal. The authors claimed that they are the first to utilize a low cost and conductive polyester filter cloth, modified by *in situ* formed PANi (polyaniline)-phytic acid (PA) as cathode. The potential anode cell was not affected by the anode types as it was regulated by the respiration of inoculated *Shewanella oneidensis*. This MBR–MFC system was able to remove COD of the synthetic wastewater up to 95% with the modified membrane as cathode. However, the electricity was indeed low and further studies are needed to improve the electricity generation.

Y.-K. Wang et al. (2013) studied the utilization of MFC in MBR to mitigate membrane fouling. The results showed that the formed electric field reduced the deposition of sludge on the membrane surface by enhancing the electrostatic repulsive force between them. The other reason that contributed to membrane fouling is the H_2O_2 produced at the cathode which is able to remove the membrane foulants. This system exhibited a good performance where the COD removal was 94%, 97% for ammoniacal nitrogen removal as well as a low effluent turbidity below 2 NTU. Liu et al. (2014) studied the integration of MBR–MFC for electricity generation, fouling mitigation, and artificial wastewater treatment. The authors identified that the sludge properties and aeration in the cathodic chamber were the main factors affecting electricity generation. They found that MFC was successfully alleviated membrane fouling under closed-circuit condition. Electrons generated in the anaerobic chamber can be transferred to cathode membranes via an external circuit and the protons generated in anode zones would cross the separator to cathode chambers (Figure 6.2).

This exerts an additional repulsion force to membrane foulants such as negatively charged sludge and organic matter and mitigates membrane fouling. This system was successful in the removal of offensive smells, TN, COD, and turbidity after aerobic treatment and filtration from the synthetic wastewater, and similar to the sequential anaerobic–aerobic system. Tian et al. (2015) studied the in situ integration of MFC with hollow-fiber MBR for wastewater treatment and membrane fouling mitigation. The removal efficiencies of COD, ammoniacal nitrogen, and TN in the system were improved by 4.4%, 1.2%, and 10.3%, respectively. This system also alleviated the membrane fouling by sludge modification where less loosely bound extracellular polymeric substances (LB-EPS), a lower SMP_p/SMP_c ratio, more homogenized sludge flocs, and less filamentous bacteria were obtained in the MFC–MBR compared to conventional MBR. Another study conducted by Su et al. (2013) investigated the treatment efficiency, sludge reduction, energy recovery, and membrane fouling in MFC–MBR. A higher sludge reduction (5.1% higher than conventional MBR), electricity recovery, and membrane fouling mitigation were obtained in the system.

In conclusion, MBR–MFC achieved better performance with potential advantages in energy consumption and recovery as well as minimization of membrane fouling. It can also be turned into a simple and rapid COD biosensor. Compared to conventional COD sensors, the bio-cathode MDC-based biosensor does not need external power sources. It makes intelligent use of the electrical energy (J. Wang et al., 2014). Some researchers utilized the membrane as cathode and filtration membrane. The others utilized the aeration in cathode chamber to enhance the reduction of electrons and mitigation of membrane fouling. Whatever the design and material of MFC–MBR are, the goal was to target low-cost sustainable development of MBR technology. Further development would address the challenges as in scaling up system, capital cost, and operation cost.

6.5 Conclusion

Many advantages associated with MBR technology make it a reliable and valuable option, more favorable over other waste management techniques. Future research in MBR is likely to focus on reduction in energy demand and membrane fouling during the operation. More and more new structures of MBR had been proposed for practical application in environmental engineering. MBR showed a good performance in terms of high organic removal and it will be an attractive alternative for water reuse and recycle in the near future. Other developments in the integrated MBR are required to address design issues for simultaneous treatment and valorization of wastewater. There are several options one can choose in order to find the most appropriate MBR technology for a particular region. The attractive advantages and interesting engineering characteristics of integrated MBR have great potential to play an important role in wastewater treatment for sustainable development. The continuous effort in academia and industry will contribute to the emergence of integrated MBR in the treatment and valorization of wastewater.

References

Adrianus, C. V. H. and Gatze, L. 1994. Anaerobic sewage treatment: A practical guide for regions with a hot climate. The University of Michigan, John Wiley & Sons, Chichester, UK.

Alizadeh Fard, M., Aminzadeh, B., Taheri, M., Farhadi, S., and Maghsoodi, M. 2013. MBR excess sludge reduction by combination of electrocoagulation and Fenton oxidation processes. *Separation and Purification Technology*, 120, 378–385, doi: http://dx.doi.org/10.1016/j.seppur.2013.10.012.

Bani-Melhem, K. and Smith, E. 2012. Grey water treatment by a continuous process of an electrocoagulation unit and a submerged membrane bioreactor system. *Chemical Engineering Journal*, 198–199, 201–210, doi: http://dx.doi.org/10.1016/j.cej.2012.05.065.

Chen, L., Gu, Y., Cao, C., Zhang, J., Ng, J.-W., and Tang, C. 2014. Performance of a submerged anaerobic membrane bioreactor with forward osmosis membrane for low-strength wastewater treatment. *Water Research*, 50, 114–123, doi: http://dx.doi.org/10.1016/j.watres.2013.12.009.

Gallucci, F., Basile, A., and Hai, F. I. 2011. Introduction—A review of membrane reactors. In: *Membranes for Membrane Reactors: Preparation, Optimization and Selection* (Eds A. Basile and F. Gallucci), John Wiley & Sons, Ltd, Chichester, UK. doi: 10.1002/9780470977569.ch

Gao, D.-W., Hu, Q., Yao, C., Ren, N.-Q., and Wu, W.-M. 2014. Integrated anaerobic fluidized-bed membrane bioreactor for domestic wastewater treatment. *Chemical Engineering Journal*, 240, 362–368, doi: http://dx.doi.org/10.1016/j.cej.2013.12.012.

Ge, Z., Ping, Q., and He, Z. 2013. Hollow-fiber membrane bioelectrochemical reactor for domestic wastewater treatment. *Journal of Chemical Technology and Biotechnology*, 88(8), 1584–1590, doi: http://dx.doi.org/10.1002/jctb.4009.

Giacobbo, A., Feron, G. L., Rodrigues, M. A. S., Ferreira, J. Z., Meneguzzi, A., and Bernardes, A. M. 2014. Integration of membrane bioreactor and advanced oxidation processes for water recovery in leather industry. *Desalination and Water Treatment*, 140, 1–10 doi: http://dx.doi.org/10.1080/19443994.2014.956346.

Goh, S., Zhang, J., Liu, Y., and Fane, A. G. 2014. Membrane distillation bioreactor (MDBR)—A lower green-house-gas (GHG) option for industrial wastewater reclamation. *Chemosphere*, doi: http://dx.doi.org/10.1016/j.chemosphere.2014.09.003.

Haandel, A. v. and Lubbe, J. 2011. Membrane bioreactor. In: *Handbook of Biological Wastewater Treatment: Design and Optimisation of Activated Sludge Systems* (Ed. 2). IWA Publishing, London, UK.

Hanft, S. March 2011. *Membrane Bioreactors: Global Markets*. Report Code: MST047C: BCC research.

Huang, J., Wang, Z., Zhu, C., Ma, J., Zhang, X., and Wu, Z. 2014. Identification of microbial communities in open and closed circuit bioelectrochemical MBRs by high-throughput 454 pyrosequencing. *PLoS ONE*, 9(4), e93842, doi: http://dx.doi.org/10.1371/journal.pone.0093842.

Jensen, P. 2015. Integrated Agri-Industrial Wastewater Treatment and Nutrient Recovery, Year 3. Australian Meat Processor Corporation, Report ID:2013-5018, North Sydney.

Keerthi, S., Vinduja, V. and Balasubramanian, N. 2013. Electrocoagulation-integrated hybrid membrane processes for the treatment of tannery wastewater. *Environmental Science and Pollution Research*, 20(10), 7441–7449, doi: http://dx.doi.org/10.1007/s11356-013-1766-y.

Laera, G., Cassano, D., Lopez, A., Pinto, A., Pollice, A., Ricco, G., and Mascolo, G. 2011. Removal of organics and degradation products from industrial wastewater by a membrane bioreactor integrated with Ozone or UV/H_2O_2 treatment. *Environmental Science and Technology*, 46(2), 1010–1018, doi: http://dx.doi.org/10.1021/es202707w.

Lamsal, R. 2012. *Advanced Oxidation Processes: Assessment of Natural Organic Matter Removal and Integration with Membrane Processes*. Dalhousie University, Halifax, Nova Scotia.

Li, J., Ge, Z., and He, Z. 2014. Advancing membrane bioelectrochemical reactor (MBER) with hollow-fiber membranes installed in the cathode compartment. *Journal of Chemical Technology and Biotechnology*, 89(9), 1330–1336, doi: http://dx.doi.org/10.1002/jctb.4206.

Li, N., Liu, L., and Yang, F. 2014. Power generation enhanced by a polyaniline–phytic acid modified filter electrode integrating microbial fuel cell with membrane bioreactor. *Separation and Purification Technology*, 132, 213–217, doi: http://dx.doi.org/10.1016/j.seppur.2014.05.028.

Liu, J., Liu, L., Gao, B., Yang, F., Crittenden, J., and Ren, N. 2014. Integration of microbial fuel cell with independent membrane cathode bioreactor for power generation, membrane fouling mitigation and wastewater treatment. *International Journal of Hydrogen Energy*, 39(31), 17865–17872, doi: http://dx.doi.org/10.1016/j.ijhydene.2014.08.123.

Logan, B. E. 2008. *Microbial Fuel Cells*. John Wiley and Sons, Inc., State College, Pennsylvania.

López, J. L. C., Reina, A. C., Gómez, E. O., Martín, M. M. B., Rodríguez, S. M., and Pérez, J. A. S. 2010. Integration of solar photocatalysis and membrane bioreactor for pesticides degradation. *Separation Science and Technology*, 45(11), 1571–1578, doi: http://dx.doi.org/10.1080/01496395.2010.487465.

Mascolo, G., Laera, G., Pollice, A., Cassano, D., Pinto, A., Salerno, C., and Lopez, A. 2010. Effective organics degradation from pharmaceutical wastewater by an integrated process including membrane bioreactor and ozonation. *Chemosphere*, 78(9), 1100–1109, doi: http://dx.doi.org/10.1016/j.chemosphere.2009.12.042.

McCabe, B. K., Hamawand, I., Harris, P., Baillie, C., and Yusaf, T. 2014. A case study for biogas generation from covered anaerobic ponds treating abattoir wastewater: Investigation of pond performance and potential biogas production. *Applied Energy*, 114, 798–808, doi: http://dx.doi.org/10.1016/j.apenergy.2013.10.020.

Merayo, N., Hermosilla, D., Blanco, L., Cortijo, L., and Blanco, Á. 2013. Assessing the application of advanced oxidation processes, and their combination with biological treatment, to effluents from pulp and paper industry. *Journal of Hazardous Materials*, 262, 420–427, doi: http://dx.doi.org/10.1016/j.jhazmat.2013.09.005.

Min, B. and Angelidaki, I. 2008. Innovative microbial fuel cell for electricity production from anaerobic reactors. *Journal of Power Sources*, 180(1), 641–647, doi: http://dx.doi.org/10.1016/j.jpowsour.2008.01.076.

Mutamim, N. S. A., Noor, Z. Z., Hassan, M. A. A., and Olsson, G. 2012. Application of membrane bioreactor technology in treating high strength industrial wastewater: A performance review. *Desalination*, 305, 1–11, doi: http://dx.doi.org/10.1016/j.desal.2012.07.033.

Mutamim, N. S. A., Noor, Z. Z., Hassan, M. A. A., Yuniarto, A., and Olsson, G. 2013. Membrane bioreactor: Applications and limitations in treating high strength industrial wastewater. *Chemical Engineering Journal*, 225, 109–119, doi: http://dx.doi.org/10.1016/j.cej.2013.02.131.

Ozgun, H., Dereli, R. K., Ersahin, M. E., Kinaci, C., Spanjers, H., and van Lier, J. B. 2013. A review of anaerobic membrane bioreactors for municipal wastewater treatment: Integration options, limitations and expectations. *Separation and Purification Technology*, 118, 89–104, doi: http://dx.doi.org/10.1016/j.seppur.2013.06.036.

Pellegrin, M.-L., Aguinaldo, J., Arabi, S., Sadler, M. E., Min, K., Liu, M., Padhye, L. P. 2013. Membrane processes. *Water Environment Research*, 85(10), 1092–1175.

Phattaranawik, J., Fane, A. G., Pasquier, A. C. S., Bing, W., and Wong, F. S. 2009. Experimental study and design of a submerged membrane distillation bioreactor. *Chemical Engineering & Technology*, 32(1), 38–44, doi: http://dx.doi.org/10.1002/ceat.200800498.

Pretel, R., Robles, A., Ruano, M. V., Seco, A., and Ferrer, J. 2014. The operating cost of an anaerobic membrane bioreactor (AnMBR) treating sulphate-rich urban wastewater. *Separation and Purification Technology*, 126, 30–38, doi: http://dx.doi.org/10.1016/j.seppur.2014.02.013.

Qu, X., Gao, W. J., Han, M. N., Chen, A., and Liao, B. Q. 2012. Integrated thermophilic submerged aerobic membrane bioreactor and electrochemical oxidation for pulp and paper effluent treatment—Towards system closure. *Bioresource Technology*, 116, 1–8, doi: http://dx.doi.org/10.1016/j.biortech.2012.04.045.

Rodríguez, F. A., Poyatos, J. M., Reboleiro-Rivas, P., Osorio, F., González-López, J., and Hontoria, E. 2011. Kinetic study and oxygen transfer efficiency evaluation using respirometric methods in a submerged membrane bioreactor using pure oxygen to supply the aerobic conditions. *Bioresource Technology*, 102(10), 6013–6018, doi: http://dx.doi.org/10.1016/j.biortech.2011.02.083.

Royan, F. 2016. Membrane multiplier: MBR set for global growth. *Water and Wastewater International*, 27.

SBI. (October 05, 2012). Global Market for Membrane Wastewater Treatment. MarketResearch.com.

Sheldon, M., Zeelie, P., and Edwards, W. 2012. Treatment of paper mill effluent using an anaerobic/aerobic hybrid side-stream Membrane Bioreactor. *Water Science and Technology*, 65(7), 1265–1272.

Skouteris, G., Hermosilla, D., López, P., Negro, C., and Blanco, Á. 2012. Anaerobic membrane bioreactors for wastewater treatment: A review. *Chemical Engineering Journal*, 198–199, 138–148, doi: http://dx.doi.org/10.1016/j.cej.2012.05.070.

Smith, A. L., Stadler, L. B., Love, N. G., Skerlos, S. J., and Raskin, L. 2012. Perspectives on anaerobic membrane bioreactor treatment of domestic wastewater: A critical review. *Bioresource Technology*, 122, 149–159, doi: http://dx.doi.org/10.1016/j.biortech.2012.04.055.

Solomou, N., Stamatoglou, A., Malamis, S., Katsou, E., Costa, C., and Loizidou, M. 2014. An integrated solution to wastewater and biodegradable organic waste management by applying anaerobic digestion and membrane bioreactor processes. *Water Practice & Technology*, 9(4), 464–474.

Su, X., Tian, Y., Sun, Z., Lu, Y., and Li, Z. 2013. Performance of a combined system of microbial fuel cell and membrane bioreactor: Wastewater treatment, sludge reduction, energy recovery and membrane fouling. *Biosensors and Bioelectronics*, 49, 92–98, doi: http://dx.doi.org/10.1016/j.bios.2013.04.005.

Tan, J.-M., Qiu, G., and Ting, Y.-P. 2015. Osmotic membrane bioreactor for municipal wastewater treatment and the effects of silver nanoparticles on system performance. *Journal of Cleaner Production*, 88, 146–151, doi: http://dx.doi.org/10.1016/j.jclepro.2014.03.037.

Tian, Y., Li, H., Li, L., Su, X., Lu, Y., Zuo, W., and Zhang, J. 2015. In-situ integration of microbial fuel cell with hollow-fiber membrane bioreactor for wastewater treatment and membrane fouling mitigation. *Biosensors and Bioelectronics*, 64, 189–195, doi: http://dx.doi.org/10.1016/j.bios.2014.08.070.

Tijing, L. D., Woo, Y. C., Choi, J.-S., Lee, S., Kim, S.-H., and Shon, H. K. 2015. Fouling and its control in membrane distillation—A review. *Journal of Membrane Science*, 475, 215–244, doi: http://dx.doi.org/10.1016/j.memsci.2014.09.042.

Torres-Sánchez, A. L., López-Cervera, S. J., de la Rosa, C., Maldonado-Vega, M., Maldonado-Santoyo, M., and Peralta-Hernández, J. M. 2014. Electrocoagulation process coupled with advance oxidation techniques to treatment of dairy industry wastewater. *International Journal of Electrochemical Science*, 9, 6103–6112.

Vijayakumar, V., Keerthi, and Balasubramanian, N. 2014. Heavy metal removal by electrocoagulation integrated membrane bioreactor. *CLEAN—Soil, Air, Water*, 532–537, doi: http://dx.doi.org/10.1002/clen.201300555.

Wang, J., Zheng, Y., Jia, H., and Zhang, H. 2014. Bioelectricity generation in an integrated system combining microbial fuel cell and tubular membrane reactor: Effects of operation parameters performing a microbial fuel cell-based biosensor for tubular membrane bioreactor. *Bioresource Technology*, 170, 483–490, doi: http://dx.doi.org/10.1016/j.biortech.2014.08.033.

Wang, P. and Chung, T.-S. 2015. Recent advances in membrane distillation processes: Membrane development, configuration design and application exploring. *Journal of Membrane Science*, 474, 39–56, doi: http://dx.doi.org/10.1016/j.memsci.2014.09.016.

Wang, Y.-K., Li, W.-W., Sheng, G.-P., Shi, B.-J., and Yu, H.-Q. 2013. In-situ utilization of generated electricity in an electrochemical membrane bioreactor to mitigate membrane fouling. *Water Research*, 47(15), 5794–5800, doi: http://dx.doi.org/10.1016/j.watres.2013.06.058.

Wang, Y.-K., Sheng, G.-P., Li, W.-W., Huang, Y.-X., Yu, Y.-Y., Zeng, R. J., and Yu, H.-Q. 2011. Development of a novel bioelectrochemical membrane reactor for wastewater treatment. *Environmental Science and Technology*, 45(21), 9256–9261, doi: http://dx.doi.org/10.1021/es2019803.

Wang, Y.-K., Sheng, G.-P., Shi, B.-J., Li, W.-W., and Yu, H.-Q. 2013. A novel electro-chemical membrane bioreactor as a potential net energy producer for sus-tainable wastewater treatment. *Scientific Reports*, 3, doi: 10.1038/srep01864 http://www.nature.com/srep/2013/130521/srep01864/abs/srep01864.html# supplementary-information.

Wang, Y.-P., Liu, X.-W., Li, W.-W., Li, F., Wang, Y.-K., Sheng, G.-P., Raymond, J. Z., Yu, H.-Q. 2012. A microbial fuel cell–membrane bioreactor integrated system for cost-effective wastewater treatment. *Applied Energy*, 98, 230–235, doi: http://dx.doi.org/10.1016/j.apenergy.2012.03.029.

Wang, Z., Huang, J., Zhu, C., Ma, J., and Wu, Z. 2013. A bioelectrochemically-assisted membrane bioreactor for simultaneous wastewater treatment and energy pro-duction. *Chemical Engineering and Technology*, 36(12), 2044–2050, doi: http://dx.doi.org/10.1002/ceat.201300322.

Wang, Z., Ma, J., Tang, C. Y., Kimura, K., Wang, Q., and Han, X. 2014. Membrane cleaning in membrane bioreactors: A review. *Journal of Membrane Science*, 468, 276–307, doi: http://dx.doi.org/10.1016/j.memsci.2014.05.060.

Wei, C.-H., Harb, M., Amy, G., Hong, P.-Y., and Leiknes, T. 2014. Sustainable organic loading rate and energy recovery potential of mesophilic anaerobic membrane bioreactor for municipal wastewater treatment. *Bioresource Technology*, 166, 326–334, doi: http://dx.doi.org/10.1016/j.biortech.2014.05.053.

Ylitervo, P., Akinbomi, J., and Taherzadeh, M. J. 2013. Membrane bioreactors' poten-tial for ethanol and biogas production: a review. *Environmental Technology*, 34(13-14), 1711–1723, doi: http://dx.doi.org/10.1080/09593330.2013.813559.

Youngsukkasem, S., Chandolias, K., and Taherzadeh, M. J. 2014. Rapid bio-methanation of syngas in a reverse membrane bioreactor: Membrane encased microorgan-isms. *Bioresource Technology*, doi: http://dx.doi.org/10.1016/j.biortech.2014.07.071.

Section III

Advanced Chemical-Physical Processes for Industrial Wastewater Treatment

7

Wet Air Oxidation Processes: A Pretreatment to Enhance the Biodegrability of Pharmaceutical Wastewater

Ee Ling Yong

CONTENTS

7.1 Introduction

Pharmaceutical compounds have been frequently detected in surface waters worldwide (Anderson et al., 2004; Zhao et al., 2009; Yoon et al., 2010; Al-Odaini et al., 2013; Komori et al., 2013; Meffe and de Bustamante, 2014). These compounds, especially endocrine disrupting compounds, have been shown to alter the sexes of aquatic organisms (Sellin et al., 2009; Yan et al., 2012; Fawell and Ong, 2012) which will eventually affect humans if this warning sign is continuously ignored. With the advancement in the biotechnology field, pharmaceutical industries have been producing new drugs each year to cater for today's demand for drugs as one of the contributing factors.

The wastewater produced by industries are highly toxic consisting of high concentration of refractory organic compounds with COD concentrations ranging from 690 to 188,108 mg/L (Cokgor et al., 2004; Arslan-Alaton and Dogruel, 2004; Suarez-Ojeda et al., 2007; Mascolo et al., 2010; Padoley et al., 2011; Wang et al., 2012; Lefebvre et al., 2014). In addition, high phosphorus and nitrogen content ranging between 3000 and 30,000 mg/L (Wang et al., 2012) and 160–7500 mg/L (Mascolo et al., 2010; Lefebvre et al., 2014) were also found in the wastewater. High nutrient content of the wastewater can lead to a eutrophication problem when released to the environment.

Although conventional biological treatment can reduce the COD of wastewater, it is not sufficient to effectively remove the remaining pharmaceutical compounds in the wastewater owing to the presence of large amounts of organic solvents (Mascolo et al., 2010) which are more readily biodegradable. Recently, the integration of a chemical treatment system, specifically wet air oxidation (WAO) processes, with the conventional biological treatment has greatly enhanced the performance of the wastewater treatment system by converting recalcitrant compounds into more readily biodegradable products (Arslan-Alaton and Dogruel, 2004; Lefebvre et al., 2014). Other chemical treatment systems such as ultraviolet (UV) irradiation are also commonly employed. However, UV irradiation treatment is relatively less effective due to presence of other stronger UV absorbers in the pharmaceutical wastewater (Arslan-Alaton and Dogruel, 2004). The Fenton reaction process was discussed in previous section, thus, the discussion in this section is limited to the WAO and ozonation processes.

7.2 Wet Air Oxidation Processes

WAO, also known as liquid phase oxidation, is an oxidation process using pure oxygen gas or air to oxidize organic or inorganic compounds at high temperatures and pressures (Mishra et al., 1995; Luck, 1999). Temperature can affect the gas–liquid state of water and oxygen gas solubility. Therefore, elevated pressures are essential to maintain water in its liquid form and to enhance oxygen solubility in order to drive the oxidation process. Typically, the temperature and pressure for wet oxidation range from 180°C to 315°C and 0.5 to 20 MPa, respectively. Under these conditions, 75%–90% of organic compounds are converted into low molecular weight by-products, such as acetic and propionic acids, methanol, ethanol, and acetaldehyde, that are easily removed by the following biological processes (Luck, 1996, 1999).

7.3 The Reaction Mechanisms

The reaction mechanisms of WAO processes are normally elucidated using either phenol or carboxylic acids (Li et al., 1991; Rivas et al., 1998). The mechanisms as shown in Equations 7.1 through 7.8 are applicable on pharmaceutical wastewater as almost all pharmaceutical compounds contain phenolic and/or carboxylic acid functional groups.

$$PhOH + O_2 \rightarrow PhO\cdot + HOO\cdot \qquad (7.1)$$

$$PhOH + HOO \cdot \rightarrow PhO \cdot + H_2O_2 \tag{7.2}$$

$$RH + O_2 \rightarrow R \cdot + HOO \cdot \tag{7.3}$$

$$RH + HOO \cdot \rightarrow R \cdot + H_2O_2 \tag{7.4}$$

$$H_2O_2 \rightarrow 2 \cdot OH \tag{7.5}$$

$$PhOH + \cdot OH \rightarrow Ph \cdot + H_2O \tag{7.6}$$

$$Ph \cdot + O_2 \leftrightarrow PhOO \cdot \tag{7.7}$$

$$Ph \cdot + PhOO \cdot \rightarrow Ph \cdot + PhOOH \tag{7.8}$$

$$PhOOH \rightarrow PhO \cdot + \cdot OH \tag{7.9}$$

$$RH + \cdot OH \rightarrow R \cdot + H_2O \tag{7.10}$$

$$R \cdot + O_2 \rightarrow ROO \cdot \tag{7.11}$$

$$ROO \cdot + RH \rightarrow ROOH + R \cdot \tag{7.12}$$

R denotes organic functional groups.

Based on Equations 7.1 through 7.5, WAO processes are initiated via the reaction of oxygen with the O–H bond and the weakest C–H bond of phenol and carboxylic acids, respectively. The reactions produce peroxyl radical which forms the hydroxyl radical, a nonselective highly reactive oxidant. This oxidant can react with any organic compounds at the rate of 10^8–10^{10}/M/s⁻ (Haag and Yao, 1992).

The series of chain reactions (Equations 7.6 through 7.12) often lead to the lower molecular weight compounds, mainly formic and acetic acids (Li et al., 1991; Sanchez-Oneto et al., 2004) which can be easily converted to carbon dioxide and water in the biological phase.

7.4 Catalytic WAO Processes

WAO processes are carried out under extreme conditions leading to the corrosion of the construction materials (Levec and Pintar, 2007). In addition, high energy consumption has become another contributing factor for industries to divert their attention to catalytic WAO (CWAO) for waste treatment purpose. The introduction of a catalyst in WAO can significantly reduce the energy consumption and temperature, yet can achieve high oxidation rates (Luck, 1999; Sanchez-Oneto et al., 2004). Two types of CWAO technologies were developed over the decades, namely heterogeneous and homogeneous CWAO. Heterogeneous CWAO catalysts were made from precious metals deposited on titania–zirconia oxides while homogeneous catalysts comprised only of precious metals (Luck, 1996). Among the two CWAOs, heterogeneous CWAO provides a better performance over homogeneous CWAO due to its capability to oxidize acetic acid and ammonia (Bhargaya et al., 2006). Therefore, the catalysts and their respective operating conditions in Table 7.1 only compile information related to recently developed heterogeneous catalysts.

TABLE 7.1

CWAO Catalysts and Their Operating Conditions

Catalysts	Operating Conditions	References
Ru/TiO_2	COD removal: >90% Pressure: 76 bar Temperature: 180°C Operating time: 480 min	Arena et al. (2014)
Activated carbon/$Ce_{0.2}Zr_{0.85}O_2$	TOC removal: 53% Pressure: 60 bar Temperature: 180°C Operating time: 150 min Activated carbon specific area loss: 80%	Heponiemi et al. (2011)
$CuO–ZrO_2–La_2O_3/$ZSM-5	COD removal: 98.7% pH 7 Pressure: 35 bar Temperature: 240°C Noncatalyst COD removal: 35.8%	Peng and Zeng (2011)
CuO/activated carbon	COD removal: >90% Temperature: 80°C Operating time: 20–30 min	Liou and Chen (2009)
Cu/carbon nanofibers	TOC removal: 74.1% Pressure: 63–87 bar Temperature: 120–160°C Operating time: 80 min	Rodriguez et al. (2008)
$CeO_2–SiO_2$	COD removal: 44%–53.8% Pressure: 20–35 bar Temperature: 157–227°C Operating time: 120 min	Goi et al. (2004)
$Pd/Al_2O_3Ru/TiO_2Ru/$Pd/TiO	COD removal: 74.9%–85.2% Total nitrogen removal: 5.6%–66.5% Temperature: 250°C Operating time: 30 min	Lei et al. (2005)

However, several challenges on the CWAO catalysts include leaching of the active ingredient, material surface area losses, and the deposition of organic as well as inorganic materials (Levec and Pintar, 2007). Those challenges could incur unnecessary expenditure during operation.

7.5 The Enhancement of Pharmaceutical Wastewater Biodegradability

Conventional biological treatment for pharmaceutical wastewater may not be applicable to pharmaceutical wastewater that contains high amount of COD. Previous studies have proven that the degradation of some organic and inorganic compounds was inhibited at the concentration as low as 20 mg/L depending on the complexity and toxicity of a compound (Hung and Pavlostathis, 1997; Bajaj et al., 2009). This constraint makes WAO process an attractive option as a pretreatment to enhance the biodegradability of pharmaceutical wastewater. Nonetheless, a treatment system coupling WAO and biological treatment can be overdesigned if the optimal pollutant concentration and wastewater flowrate are not known (Collado et al., 2012). For example, a single WAO process is sufficient for a low flowrate but concentrated pollutants. The free radical chain reactions in WAO at this point not only effectively remove the pollutants but also omit the necessity in managing the sludge generated from the biological process. On the contrary, wastewater with high flowrate and pollutant concentration is just too expensive to run on either WAO or biological process alone. Therefore, combination of both enhances the treatability of a high strength wastewater as well as reduces potential sludge generation. The treatment performance and operating conditions of the noncatalytic and CWAO are summarized in Table 7.2.

From Table 7.2, the absence of a catalyst in a WAO system can only remove at most 38% COD whereas the presence of a catalyst can achieve a removal efficiency of 99.5%. This denotes that a WAO system has considerably lower oxidation capacity compared to a CWAO treatment system. Furthermore, the pH of the wastewater played an important role on the removal efficiency. As pH affects the free radical chain reactions, the oxidation capacity in the (C)WAO system is highly pH dependent. An increase in pH decreased the COD removal (Zhan et al., 2013) and pH of the wastewater fluctuated during the treatment process. For pH above 7, pH was reduced while those below 7 the pH was found to increase (Hosseini et al., 2011, 2012; Lei et al., 2011). Increase in temperature and pressure also accelerated the removal process (Zhan et al., 2013).

In general, organics in pharmaceutical wastewater are comprised of complex structure compounds. Upon (C)WAO process, short-chain carboxylic acids such as acetic and propionic acids that are susceptible to biodegradation

TABLE 7.2

Noncatalytic and Catalytic WAO Operating Conditions and Treatment Efficiencies on Pharmaceutical Wastewater

With or Without Catalyst (YES/NO)	Type of Catalysts	Pharmaceutical Wastewater Characteristics	Operating Conditions	Removal Efficiencies
NO		pH = 6.9 COD = 124,500 mg/L DOC = 40,200 mg/L	Pressure = 33–98 bar Temperature = 240–280°C	COD = 63%–80%
YES and NO	Ru/Ir monolith	pH = 1–14 COD = 10,788–751,800 mg/L	Pressure = 50 bar Temperature = 230–250°C Operation time = 3–14.5 h	COD = 25%–98%
YES	MnCeO$_2$	pH = 2.62 COD = 8000–12,000 mg/L BOD$_5$ = 500–1000 mg/L	Pressure = 5–10 bar Temperature = 160–220°C Operation time = 20–80 h	COD = 56%–80%
YES	Ru/Ir oxide-coated Ti monolith	pH = 13–14 COD = 35,637 mg/L	Pressure = 50 bar Temperature = 200–230°C Operation time = 60–300 min	COD = 38% (no catalyst) COD = 59%–97% (with catalysts)
YES	Fe^{2+}	COD = 9016 mg/L Salicylic acid (SA) = 1464 mg/L Phenol = 1996 mg/L 5-Hydroxyisophthalic acid (5HIA) = 122.7 mg/L	Pressure = 10.1 bar Temperature = 140°C Operation time = 1.6–4 h	COD > 95% SA = 66% Phenol = 99.5% 5HIA = 99.2% mg/L
YES	Fe$_2$O$_3$/SBA-15	pH = 5.6 COD = 1901 mg/L	Pressure = 50 bar Temperature = 70–80°C Operation time = 200 min	TOC mineralization = 55%–60%
YES	Cu^{2+} and polyoxometalates	COD = 3201 mg/L TOC = 1470 mg/L	Pressure = 14.1 bar Temperature = 250°C Operation time = 60 min	COD and TOC removal are >40%
YES	Ru/TiO$_2$	pH = 11.4–12.2 COD = 23,900–44,300 mg/L	Pressure = 10–100 bar Temperature = 170–300°C Operation time = 60–90 min	COD >90%

were observed (Mishra et al., 1995; Melero et al., 2009; Hosseini et al., 2011, 2012; Wang et al., 2012; Sergio et al., 2013). Thus, following a biological treatment after a (C)WAO process can completely mineralize the complex and recalcitrant organic compounds. However, it is important to note that the fluctuation in pollutant loading can affect biodegradation as microbes require time to adapt to the new pollutant loading.

References

Al-Odaini, N.A., M.P. Zakaria, M.I. Yaziz, S. Surif, N. Kannan 2013. Occurrence of synthetic hormones in sewage effluents and Langat river and its tributaries, Malaysia. *International Journal of Environmental Analytical Chemistry*. 93 (14): 1457–1469.

Anderson, P.D., V.J. D'Aco, P. Shanahan, S.C. Chapra, M.E. Buzby, V.L. Cunningham et al. 2004. Screening analysis of human pharmaceutical compounds in US surface waters. *Environmental Science and Technology*. 38 (3): 838–849.

Arena, F., C. Italiano, G.D. Ferrante, G. Trunfio, L. Sparado 2014. A mechanistic assessment of the wet air oxidation activity of $MnCeO_x$ catalyst toward toxic and refractory organic pollutants. *Applied Catalysis B: Environmental*. 144: 292–299.

Arslan-Alaton, I., S. Dogruel 2004. Pre-treatment of penicillin formulation effluent by advanced oxidation processes. *Journal of Hazardous Materials*. 112 (1–2): 105–113.

Bajaj, M., C. Gallert, J. Winter 2009. Phenol degradation kinetics of an aerobic mixed culture. *Biochemical Engineering Journal*. 46 (2): 205–209.

Bhargaya, S.K., J. Tardio, J. Prasad, K. Foger, D.B. Akolekar, S.C. Grocott 2006. Wet oxidation and catalytic wet oxidation. *Industrial and Engineering Chemistry Research*. 45 (4): 1221–1258.

Cokgor, E.U., L.A. Alaton, O. Karahan, S. Dogruel, D. Orhon 2004. Biological treatability of raw and ozonated penicillin formulation effluent. *Journal of Hazardous Materials*. 116: 159–166.

Collado, S., A. Laca, M. Diaz 2012. Decision criteria for the selection of wet oxidation and conventional biological treatment. *Journal of Environmental Management*. 102: 65–70.

Fawell, J., C.N. Ong 2012. Emerging contaminants and the implications for drinking water. *International Journal of Water Resources Development*. 28 (2): 247–263.

Goi, D., C. De Leitenburg, A. Trovarelli, G. Dolcetti 2004. Catalytic wet-oxidation of a mixed liquid waste: COD and AOX abatement. *Environmental Technology*. 25 (12): 1397–1403.

Haag, W.R., C.C.D. Yao 1992. Rate constants for reaction of hydroxyl radicals with several drinking water contaminants. *Environmental Science and Technology*. 26 (5): 1005–1013.

Heponiemi, A., L. Rahikka, U. Lassi, T. Kuokkanen 2011. Catalytic oxidation of industrial wastewater under mild conditions. *Topics in Catalysis*. 54 (16–18): 1034–1041.

Hung, C.H., S.G. Pavlostathis 1997. Aerobic biodegradation of thiocyanate. *Water Research*. 31 (11): 2761–2770.

Hosseini, A.M., A. Tungler, V. Bakos 2011. Wet oxidation properties of process waste-waters of fine chemical and pharmaceutical origin. *Reaction Kinetics Mechanisms and Catalysis*. 103 (2): 251–260.

Hosseini, A.M., A. Tungler, Z. Schay, S. Szabó, J. Kristóf, E. Széles, L. Szentmiklósi 2012. Comparison of precious metal oxide/titanium monolith catalysts in wet oxidation of wastewaters. *Applied Catalysis B: Environmental*. 127: 99–104.

Komori, K., Y. Suzuki, M. Minamiyama, A. Harada 2013. Occurrence of selected pharmaceuticals in river water in Japan and assessment of their environmental risk. *Environmental Monitoring and Assessment*. 185 (6): 4529–4536.

Lefebvre, O., X. Shi, C.H. Wu, H.Y. Ng 2014. Biological treatment of pharmaceutical wastewater from the antibiotics industry. *Water Science and Technology*. 69 (4): 855–861.

Lei, Y.J., S.D. Zhang, J.C. He, J.C. Wu, Y. Yang 2005. Ruthenium catalyst for treatment of water containing concentrated of organic waste. *Platinum Metals Review*. 49 (2): 91–97.

Lei, Y.J., X.B. Wang, C. Song, F.H. Li, X.R. Wang 2011. A study on ruthenium-based catalysts for pharmaceutical wastewater treatment. *Water Science and Technology*. 64 (1): 117–121.

Levec, J., A. Pintar. 2007. Catalytic wet-air oxidation processes: A Review. *Catalysis Today*. 124: 172–184.

Li, L., P. Chen, E. F. Gloyna. 1991. Generalized kinetic model for wet oxidation of organic compounds. *AIChE Journal*. 37 (II): 1687–1697.

Liou, R.M., S.H. Chen 2009. CuO impregnated activated carbon for catalytic wet peroxide oxidation of phenol. *Journal of Hazardous Materials*. 172 (1): 498–506.

Luck, F. 1996. A review of industrial catalytic wet air oxidation processes. *Catalysis Today*. 27 (1–2): 195–202.

Luck, F. 1999. Wet air oxidation: Past, present and future. *Catalysis Today*. 53 (1): 81–91.

Mascolo, G., L. Balest, D. Cassano 2010. Biodegradability of pharmaceutical industrial wastewater and formation of recalcitrant organic compounds during aerobic biological treatment. *Bioresource Technology*. 101 (8): 2585–2591.

Meffe, R., I. de Bustamante 2014. Emerging organic contaminants in surface water and groundwater: A first overview of the situation in Italy. *Science of the Total Environment*. 481: 280–295.

Melero, J.A., F. Martinez, J.A. Botas, R. Molina, M.I. Pariente 2009. Heterogeneous catalytic wet peroxide oxidation systems for the treatment of an industrial pharmaceutical wastewater. *Water Research*. 43 (16): 4010–4018.

Mishra, V.S., V.V. Mahajani, J.B. Joshi 1995. Wet air oxidation. *Industrial & Engineering Chemistry Research*. 34 (1): 2–48.

Padoley, K.V., S.N. Mudliar, S.K. Banerjee, S.C. Deshmukh, R.A. Pandey 2011. Fenton oxidation: A pretreatment option for improved biological treatment of pyridine and 3-cyanopyridin plant wastewater. *Chemical Engineering Journal*. 166: 1–9.

Peng, Q.L., J. Zeng. 2011. Study on catalyst of catalytic wet oxidation of high-concentration industrial organic wastewater. *Advanced Materials Research*. 233–235: 2994–2999.

Rivas, F.J., S.T. Kolaczkowski, F.J. Beltran, D.B. McLurgh 1998. Development of a model for the wet air oxidation of phenol based on a free-radical mechanism. *Chemical Engineering Science*. 53 (14): 2575–2586.

Rodriguez, A., G. Ovejero, M.D. Romero, C. Diaz, M. Barreiro, J. Garcia 2008. Catalytic wet air oxidation of textile industrial wastewater using metal supported on carbon nanofibers. *Journal of Supercritical Fluids*. 46 (2): 163–172.

Sanchez-Oneto, J., J.R. Portela, E. Nebot-Sanz, E.J. Martínez de la Ossa 2004. Wet air oxidation of long-chain carboxylic acids. *The Chemical Engineering Journal.* 100 (1): 43–50.

Sellin, M.K., D.D. Snow, D.L. Akerly, A.S. Kolok 2009. Estrogenic compounds downstream from three small cities in Eastern Nebraska: Occurrence and biological effect. *Journal of the American Water Resources Association.* 45: 1.

Suarez-Ojeda, M.E., A. Guisasola, J.A. Baeza, A. Fabregat, F. Stüber, A. Fortunny, J. Font, J. Carrera 2007. Integrated catalytic wet air oxidation and aerobic biological treatment in municipal WWTP of a high-strength O-cresol wastewater. *Chemosphere.* 66: 2096–2105.

Sergio, C., Q. David, L. Adriana, D. Mario 2013. Efficiency and sensitivity of the wet oxidation/biological steps in coupled pharmaceutical wastewater treatment. *Chemical Engineering Journal.* 234: 484–490.

Wang, G.W., D. Wang, X.C. Xu, L.F. Liu, F.L. Yang 2012. Wet air oxidation of pretreatment of pharmaceutical wastewater by Cu^{2+} and $[P_xW_mO_y](q^-)$ co-catalyst system. *Journal of Hazardous Materials.* 217: 366–373.

Yan, Z.H., G.H. Lu, J.C. Liu, S.G. Jin 2012. An integrated assessment of estrogenic contamination and feminization risk in fish in Taihu Lake, China. *Ecological and Environmental Safety.* 84:334–340.

Yoon, Y., J. Ryu, J. Oh, B.G. Choi, S.A. Snyder 2010. Occurrence of endocrine disrupting compounds, pharmaceuticals and personal care products in the Han River (Seoul, South Korea). *Science of the Total Environment.* 408 (3): 636–643.

Zhan, W., X. Wang, D. Li, Y. Ren, D. Liu, J. Kang 2013. Catalytic wet air oxidation of high concentration pharmaceutical wastewater. *Water Science and Technology.* 67 (10): 2281–2286.

Zhao J.L., G.G. Ying, L. Wang, J.F. Yang, X.B. Yang, L.H. Yang, X. Li 2009. Determination of phenolic endocrine disrupting chemicals and acidic pharmaceuticals in surface water of the Pearl Rivers in South China by gas chromatography–negative chemical ionization–mass spectrometry. *Science of the Total Environment.* 407 (2): 962–974.

8

Application of Nonthermal Plasma in the Treatment of Volatile Organic Compounds from Wastewater

Abdullahi Mohammed Evuti, Mohd Ariffin Abu Hassan, Zainura Zainon Noor, and Raja Kamarulzaman Raja Ibrahim

CONTENTS

8.1 Introduction

Organic compounds are carbon-based chemical compounds, which include a wide range of hydrocarbons, halocarbons, oxygenated hydrocarbons, organic nitrates, and organic sulfides. Volatile organic compounds (VOCs) have been given different definitions by different organizations and in different fields of interest. National pollution inventory defined VOCs as any compound based on a carbon chain or rings (and also containing hydrogen) with a vapor pressure greater than 0.01 kPa at 20°C, excluding methane, carbon dioxide, carbon monoxide, carbonic acids, carbonate salts, and metallic carbides. These compounds may also contain oxygen, nitrogen, and other elements (NPI, 2009).

The U.S. Environmental Protection Agency (USEPA) for regulatory purposes gave a broad definition of VOCs as any volatile compound of carbon except those specifically exempted (USEPA, 2011). Those exempted are nonreactive and slow reactive VOCs. However, VOCs may also be defined as organic chemicals with high vapor pressure at ordinary, room temperature conditions. This is due to their low boiling points, which causes large numbers of molecules to evaporate from the liquid or solid form of the compound into the surrounding air. For instance, formaldehyde with a low boiling point of –19°C (–2°F), slowly leaves the paint and gets into the air. Some common examples include acetone, benzene, ethylene glycol, formaldehyde, methylene chloride, perchloroethylene, toluene, xylene, and 1,3-butadiene (Goldstein and Galbally, 2007; Zorgoski et al., 2006).

8.2 Sources of VOCs

Many VOCs are human-made (anthropogenic) chemicals that are used or produced in the manufacturing of paints, adhesives, petroleum products, pharmaceuticals, and refrigerants. The dominant anthropogenic VOC sources are vehicular and industrial emissions from fossil fuel combustion, liquefied petroleum gas (LPG) leakages, fuel evaporation, petroleum distillation, and industrial solvents. The level of emission increases with increasing traffic densities and the ambient concentration is also affected by the nature of fuels used, type and age of vehicles, flow rates and speed of traffic as well as environmental conditions of the city (Elbir et al., 2006; Sanchez et al., 2008). Many are also compounds of fuels, solvents, hydraulic fluids, paint thinners, and dry cleaning agents commonly used in urban settings such as bleach (Zorgoski et al., 2006). Some of them are also produced by plants, animals, microbes, and fungi (biological or biogenic).

Biological sources emit an estimated 1150 teragrams of carbon per year in the form of VOCs (Goldstein and Galbally, 2007). Plants produce a majority of the VOCs with isoprene as the main compound. Other biological sources are animals, microbes, and fungi such as molds. VOC emissions in plants occur from the leaves through the stomata and it is affected by two factors; temperature, as it determines the rate of volatilization and growth and, sunlight, as it determines the rate of biosynthesis.

8.3 VOC Treatment Methods

The treatment of VOCs from wastewater can be accomplished either directly such as direct destruction of methanol in wastewater in a trickle-bed reactor packed with a hydrophobic catalyst operated at a low temperature (60–80°C) near atmospheric pressure or indirectly by separation from the wastewater which is then followed by another treatment process such as thermal destruction or recovery (Abdullahi et al., 2013a; Chuang et al., 1992). The conventional methods of treating VOCs include air stripping, adsorption, membrane technology, electrochemical technology, anaerobic/aerobic biological treatments, and bioreactor (Mourad et al., 2012; Navaladian et al., 2007; Schultz, 2005). Other methods include advanced oxidation processes and plasma technology, and integrated (hybrid) systems (Abdullahi et al., 2013b; Chuang et al., 1992; Hammer, 1999; Koutsospyros et al., 2004; Li et al., 2011).

8.4 Nonthermal Plasma Application in the Treatment of VOCs from Wastewater Plasma and Plasma Generation

Plasma can be defined as an ionized gaseous state of matter referred to as the *fourth state* of matter (Nehra et al., 2008). As the temperature increases and the molecules become more energetic, this leads to a phase change from solid, liquid, gas, and finally to plasma. This is illustrated in Figure 8.1.

Plasma is therefore an electrified gas with a chemically reactive media consisting of electrons, positive and negative ions, free radicals, metastables, gas atoms, and molecules in the ground or excited states (Fridman, 2008; Gomez et al., 2009). The electrons and ions give plasma electrical and electromagnetic properties. Where there is equal numbers of positive ions and negative electrons in the gas, it is referred to a *quasi-neutral* state (Nehra et al., 2008). However, in some circumstances not all the particles in plasma may be ionized. The ratio of major charged species to that of neutral gas is referred to

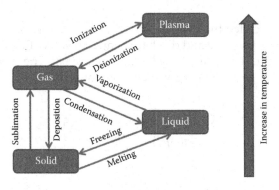

FIGURE 8.1
Phase change in states of matter.

as the *ionization degree*. Completely ionized plasma has an ionization degree close to unity while for weakly ionized plasma the ionization degree is low (Fridman, 2008).

The major source of plasma in laboratory conditions is *gas discharge*. An electric discharge can be produced by applying an electric potential between two electrodes placed in a glass tube filled with various gases or evacuated (Fridman, 2008; Nehra et al., 2008). Plasma is produced when the rate at which electrons are produced due to the application of high voltage (ionization) is more than the rate of electron loss (electron-ion recombination). The lowest voltage to initiate plasma is termed as the breakdown voltage (V_b). Therefore, generation of plasma will be sustained only if the applied voltage is greater than the breakdown voltage. This criterion for sustaining discharge was expressed mathematically as Equations 8.1 and 8.2 and called the Townsend breakdown mechanism and γ and α are the Townsend ionization coefficients (Fridman, 2008).

$$1 - \gamma(e^{\alpha d} - 1) = 0 \tag{8.1}$$

$$e^{\alpha d} = \left(1 = \frac{1}{\gamma}\right) \tag{8.2}$$

The second criterion for sustaining discharge is the Paschen breakdown voltage criteria. From the result of the research carried out by Paschen in 1889, he found out that for a fixed set of gas compositions, electrode configuration, and electrode material, the breakdown voltage is only a function of the product of pressure (p) and the discharge gap (d). Expressed mathematically as

$$V_b = f(pd) \tag{8.3}$$

Equation 8.3 is called the Paschen law. And the plot of breakdown voltage versus product of pressure and discharge gap (pd) value (Figure 8.2) is called the Paschen curve (Fridman, 2008; Nehra et al., 2008). The breakdown voltage in gas discharge plasma is given as in Equation 8.4

$$V_b = \frac{Bpd}{In(Apd) - In\left[In\left(1 + \frac{1}{\gamma}\right)\right]} \tag{8.4}$$

where A and B are constants

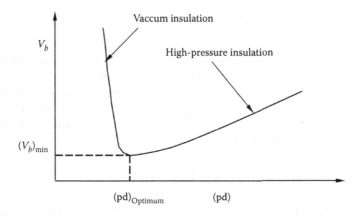

FIGURE 8.2
Paschen curve for breakdown voltage versus *pd*. (Adapted from Nehra, V. et al., 2008. *International Journal of Engineering.* 2(1), 53–68.)

8.5 Mechanism of Plasma Decomposition of VOCs

The plasma-chemical process involves a large number of elementary reactions. The sequence of changes of initial chemical substances and electric energy into products and thermal energy is referred to as the mechanism of the plasma-chemical process. The key process in plasma is ionization, which is the conversion of neutral atoms or molecules into electrons and positive ions. Ionization may occur through the following ways.

1. Direct ionization by electron impact
2. Stepwise ionization by electron impact
3. Ionization by collision of heavy particles
4. Photo-ionization
5. Surface ionization (electron emission)

Consider the collision of a single electron, e with a gas atom M or gas molecule MN, the excitation, ionization, dissociation, and charge transfer process can be illustrated by Equations 8.5 through 8.9 (Nehra et al., 2008).

$$\text{Excitation} \quad e^- + M \rightarrow M^* + e^- \tag{8.5}$$

$$\text{Ionization} \quad e^- + M \rightarrow M^+ + e^- + e^- \tag{8.6}$$

$$\text{Dissociation} \quad e^- + MN \rightarrow M + N + e^- \tag{8.7}$$

$$\text{Charge Transfer} \quad M^+ + N \rightarrow M + N^+ \tag{8.8}$$

The electron and excited molecules collide with other gas molecules and then induce secondary reactions. These include radical–radical, ion–ion, and radical–neutral atom or molecules recombination reactions, usually within a time frame of 10^{-8}–10^{-2} s. For example, the exited state of M^* has energy to cause reaction with molecule N in the reactor to produce products/by-products J and K as illustrated below

$$M^* + N \rightarrow J + K \tag{8.9}$$

The processes illustrated by Equations 8.5 through 8.9 are dependent on the voltage applied to the electrons. This shows the significant role of the power source and method for energizing the electrons. However, the over-all efficiency of the plasma process is also influenced by the type of secondary reactions, which involve various elemental processes leading to the formation of various by-products. In short, plasma chemistry can be broadly divided into primary process and secondary process. The primary process is mainly electron impact reactions which lead to the generation of active species such as ion radicals and metastables. Metastables are also important active species with high threshold energy and long life time. Chen et al. (2006), reported that a metastable of nitrogen, $N_2(A_3\Sigma_u^+)$, with a threshold energy and a lifetime of 6.17 eV and 2 s, has been confirmed to be capable of decomposing gaseous pollutants and producing the seed electrons necessary for the formation of atmospheric pressure glow discharge (APGD). The secondary process is induced by the reaction products from primary process. The timescales for the different reactions in plasma displayed in Figure 8.3 clearly shows that the timescale of the reactions in primary process (10^{-10}–10^{-8} s) are much shorter than that of the secondary process (10^{-7}–10^{-2} s).

8.6 Air Stripping: Nonthermal Plasma System for the Treatment of Toluene and Xylene Wastewater

Though electrohydraulic discharge has been successfully applied in the decomposition of VOCs in water (Locke et al., 2006; Malik, 2010), the high density of liquid prevents electrons from accelerating and undergoing dis-sociative collision, unless the electric field is in several orders of magnitude higher compared to that of plasma's at atmospheric pressure, therefore, electron avalanches are almost impossible inside a liquid because of low mobility and high recombination rate (Locke et al., 2006; Vandenbroucke

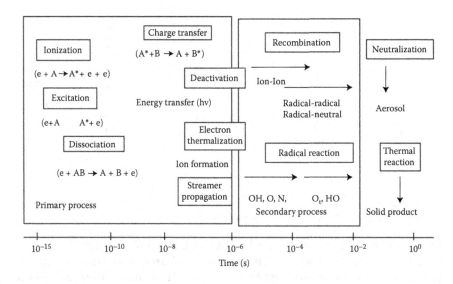

FIGURE 8.3
Timescale for different reactions in plasma. (Adapted from Kim, H. H. et al., 2002. *Journal of Electrostatics.* 55, 25–41.)

et al., 2011a,b). Also, there are reports of some cases of harmful by-products formation due to incomplete destruction of the VOCs (Anders et al., 1997; Creyghton, 1997). An integrated process therefore serves as an alternative VOC treatment method. Treatment of toluene and xylene from wastewater was studied using integrated air stripping and ferroelectric packed-bed nonthermal plasma (NTP) reactor system.

8.7 Materials and Methods

Toluene and xylene were obtained from Merck Sdn Bhd. Malaysia with greater than 99.5% purity. Synthetic wastewater containing 1500 ppm of toluene and xylene were prepared. The experimental set up is shown in Figure 8.4. It consists of a custom-made pilot scale packed with column air stripper (Model 2T4H) from Branch Environmental Corporation USA made of a 1.5 m stainless steel tube of 0.05 m internal diameter filled with 6 mm ceramic Raschig rings packing. The height of the packing is 1.15 m which is equivalent to a packing volume of $2.26 \times 10^{-3} \text{ m}^3$. The ferroelectric packed-bed NTP reactor consists of a Pyrex glass tube of 1 inch internal diameter and standard length and 3 mm diameter barium titanate ($BaTiO_3$) pellets (dielectric constant of 10,000) from Fuji Titanium Industry Co., Ltd, Japan as dielectric material packed in between the two electrodes made from

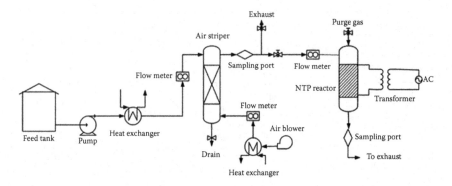

FIGURE 8.4
Integrated air and nonthermal plasma reactor system.

stainless steel. The electrodes and BaTiO$_3$ pellets were supported with two funnel head stainless steel tubes. O-rings smeared with vacuum grease were used to fix the packed bed in between the two mesh electrodes and to ensure that it was properly sealed. Teflon caps equipped with O-rings were then used to seal both ends of the Pyrex tube to create a properly confined area in the Pyrex tube. High voltage electric connectors were fixed to the two ends of the stainless steel tube. The plasma reactor was powered with a 20 kV/3 A Ac power supply transformer with input power of 240 V and 50 Hz frequency.

To remove the toluene and xylene from the wastewater using the air stripper, the airflow rate (7.12 L min^{-1}) was set using a rotameter and the wastewater inlet was also set to 0.12 L min^{-1} by adjusting the rotameter while the wastewater and air heaters were set to 50°C. The air and contaminated water were then pumped into the air stripper in counter-current operation. The treated wastewater was collected at the bottom while VOC rich air which comes out at the top of the column was sent to the nonthermal plasma reactor. The decomposition of toluene and xylene in the ferroelectric packed-bed NTP was conducted at applied voltages of 12.32, 13, 14, 15, and 15.68 kV and fixed discharge gap and flow rate of 25 mm and 3.54 L min^{-1}, respectively. A high voltage probe was used to measure the applied voltage and the current was measured using a digital Pico scope.

The concentrations of toluene and xylene in the outlet gas from the air stripper were first determined with a frontier FT-IR spectrometer (Perkin Elmer) coupled with a cyclone gas cell accessory (cyclone™ C5) manufactured by Specac Ltd. The gas streams were passed into multiple pass optical gas cell (Specac Ltd.) with a white-type mirror arrangement with an optical path length set at 8 m. Sample spectrum for each operating condition was captured at room temperature and atmospheric pressure at a spectra resolution of 1 cm^{-1}. This was then followed by the determination of decomposition efficiency of toluene and xylene and by-products from nonthermal plasma treatment by passing gas streams from nonthermal plasma through the FTIR as explained earlier. Concentrations of toluene in each spectrum were determined by integrating the area under the peaks using Perkin Elmer spectrum standard v10.4.0 software and then compared with standard spectra produced by Pacific Northwest National Laboratories.

The destruction efficiency of toluene was calculated as follows:

$$\eta(\%) = \frac{[toluene]_{inlet} - [toluene]_{outlet}}{[toluene]_{inlet}} \times 100\% \tag{8.10}$$

8.8 Results and Discussions

8.8.1 Toluene and Xylene Removal from Waste Wastewater Using Air Stripper

Calculation of the toluene and xylene gas concentration was carried out by comparing the integrated area of the recorded FTIR gas absorption band for a particular gas species, INT_{exp} to the integration area of the standard sample of that species, INT_{std} with an appropriate wavelength range. The integration of the area under the selected peaks were done using Perkin Elmer spectrum standard v10.4.0 software and then the result was compared to the standard spectra produced by Infrared Analysis, In. and Pacific Northwest National Laboratories (Figures 8.5 and 8.6). It is also assumed that there are no interfering species contributing to absorbance in the spectral window. The integration was done from the baseline of the FTIR spectrum of the selected wavelength range. The baseline is defined as a straight line between the intensities at the integration limits. The gas concentration from the integrated area of the gas absorption band was then calculated by using the equation below (Raja Ibrahim, 2012).

$$N_{ppm} = \frac{INT_{exp} \times N_{std} \times L_{std}}{INT_{std} \times L_{meas}} \tag{8.11}$$

where N_{std}, L_{std} (given as 1 m), and L_{meas} in Equation 8.11 represent the concentration of the standard gas, optical path length of the standard gas measurement, and optical path length of the measured spectrum, respectively. The N_{std} for toluene and xylene were given as 100 ppm. The summary of the results is shown in Table 8.1.

The percentage removal efficiencies after decomposition of toluene and xylene in a ferroelectric packed-bed NTP reactor at applied voltages of 12.32, 13, 14, 15, and 15.68 kV, and fixed discharge gap and flow rate of 25 mm and

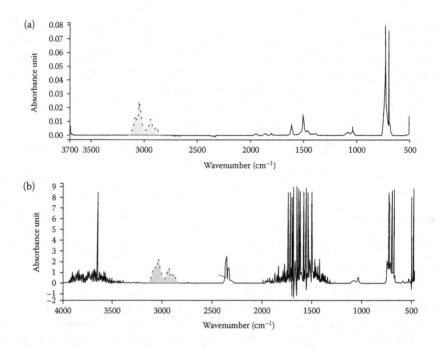

FIGURE 8.5
(a) FTIR spectrum of 100 ppm standard toluene (b) FTIR spectrum of toluene concentration in the air stream from air stripper operated at air–water ratio of 60 and column temperature of 50°C.

3.54 L min^{-1}, respectively, were calculated using Equation 8.10 and the results are shown in Table 8.1.

It can be observed from Table 8.2 that increasing the applied voltage at fixed airflow rate and discharge gap results in increased performance of the nonthermal plasma reactor for toluene and xylene removal. This can be attributed to the increased intensity of the plasma discharge observed as the applied voltage increased (Liang et al., 2009; Subrahmanyam, 2009; Vandenbroucke et al., 2011a,b). Chen et al. (2008) reported that the electron density increases with discharge current and voltage. In addition, the electric field and mean electron energy gets higher as the applied voltage is raised. The effects of applied voltage can be further understood from the FTIR spectra for toluene and xylene in the gas sample before and after plasma treatment at 15.68, 14, and 12.32 kV using a fixed discharge gap of 2.5 cm and flow rate of 3.54 L min^{-1}, respectively, as shown in Figures 8.7 and 8.8. The residual toluene and xylene after treatment in NTP reactor decreases as the applied voltage is increased from 12.32 to 15.68 kV as seen from the spectra peaks.

The removal efficiency decreased from 92% to 59.827% as the applied voltage was reduced from 15.68 to 12.32 kV for toluene. A decrease from 86.89% to 57.20% was observed for xylene under the same conditions. Chen et al.

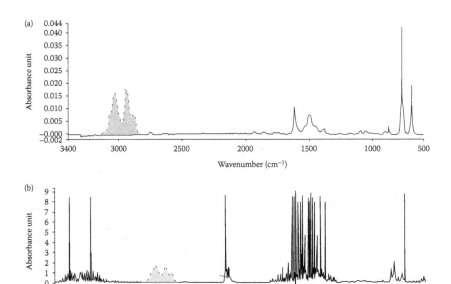

FIGURE 8.6
(a) FTIR spectrum of 100 ppm standard xylene (b) FTIR spectrum of xylene concentration in the air stream from air stripper operated at air–water ratio of 60 and column temperature of 50°C.

TABLE 8.1

Summary of the Toluene and Xylene Gas Concentrations Extracted from FTIR Measurements before Plasma Treatment

Species	Integration Range (cm^{-1})	INT_{exp}	INT_{std}	Concentration (ppm)	Percentage Removal
Toluene	3145.4–2824.3	258.63	2.16	1496.70	99.7
Xylene	3162.9–2811.0	237.94	2.24	1327.79	88.47

TABLE 8.2

Percentage Removal Efficiencies after Decomposition of Toluene and Xylene in NTP Reactor

Applied Voltage (kV)	Concentration before Plasma Treatment (ppm)		Concentration after Plasma Treatment (ppm)		Removal Efficiency	
	Toluene	Xylene	Toluene	Xylene	Toluene	Xylene
12.32	1496.70	1327.79	601.45	568.25	64.47	57.20
13.0	1496.70	1327.79	520.54	485.10	65.2	63.5
14.0	1496.70	1327.79	480.84	447.27	71.70	66.32
15.0	1496.70	1327.79	281.94	289.30	81.2	78.2
15.68	1496.70	1327.79	20.58	174.05	93.27	86.89

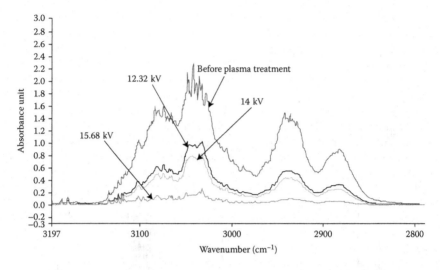

FIGURE 8.7
FTIR spectrum of toluene decomposition at fixed discharge gap of 2.5 cm and flow rate of 3.54 L min^{-1}.

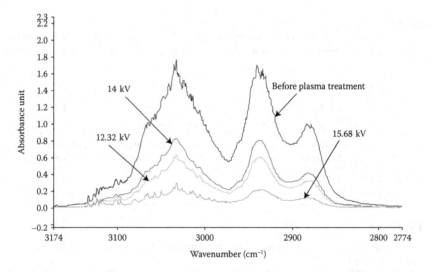

FIGURE 8.8
FTIR spectrum of xylene decomposition at fixed discharge gap of 2.5 cm and flow rate of 3.54 L min^{-1}.

(2008) and Vandenbroucke et al. (2011a,b) reported that the electron density increases with discharged current and voltage. In addition, the electric field and mean electron energy gets higher as the applied voltage is raised. The above-mentioned phenomenon is referred to as electric amplification (AF) and is defined as the ratio of the maximum electric field in the packed bed to the electric field that would be present without any packed bed (Chuang et al., 2000). Moreover, Yu-fang et al. (2006) explained that when the applied voltage is higher, the electric field is stronger and the power is higher. More electrons will therefore react with toluene and xylene and break the bond between the molecules, thus the removal efficiency increases. According to Jarrige and Vervisch (2006), the energy per pulse delivered to the plasma reactor is generally calculated by multiplying the current and voltage wave formed and integrated over the pulse width. This can be expressed mathematically as in Equation 8.12.

$$E = \int VIdt \qquad (8.12)$$

These results are also in concordance with those obtained by Li et al. (2007) for the decomposition and removal of benzene, toluene, xylene, and form-aldehyde in a positive DC streamer discharge plasma reactor. In another research by Yu-fang et al. (2006) using a wire plate dielectric barrier discharged reactor, the decomposition of toluene was also found to increase with applied voltage. Jarrige and Vervisch (2006) also explained that the energy deposited in a reactor increased with applied voltage and observed that the decomposition of propane, propene, and isopropyl alcohol increased with increased specific input energy.

8.8.2 By-Products of Toluene and Xylene Decomposition

From the zoomed recorded spectrum of the decomposition of toluene in the ferroelectric packed NTP reactor shown in Figure 8.9, four gaseous by-products (CO, CO_2, H_2O, and N_2O) were detected. This includes the presence of a residual toluene absorption band (3152–2829 cm^{-1}) with identified center of 2975 cm^{-1} since 100% decomposition was not achieved and a broad water absorption band from wavelength of 4000–3500 and 2000–1271.6 cm^{-1}. Others are CO_2 absorption band from 2396.2 to 2283.3 cm^{-1}, CO absorption band from 2050 to 2240 cm^{-1} with identified center of 2143 cm^{-1}, and N_2O absorption band from 2170 to 2260 cm^{-1} with identified center of 2226 cm^{-1} as shown in a zoomed spectrum in Figure 8.10.

Figure 8.11 is the FTIR spectrum showing the by-product of the decomposition of xylene in the NTP reactor. Similar to toluene, the presence of broad water absorption band from wavelength of 4000–3500 and 2000–1271.6 cm^{-1} was also observed in the detection of xylene decomposition byproducts. The figure also shows the presence of residual xylene absorption band (3162.9–2811 cm^{-1}) with identified center of 2972 cm^{-1} since 100% decomposition was not achieved. Others include a CO_2 absorption band from 2396.2 to 2283.3 cm^{-1} and CO absorption band with identified center of 2143 cm^{-1} and N_2O absorption band with identified center of 2226 cm^{-1}.

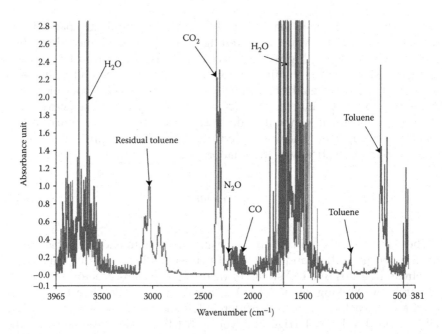

FIGURE 8.9
Spectrum of decomposition of toluene in the NTP reactor showing the by-products at applied voltages of 12.32 kV, discharge gap of 25 mm, and flow rate of 3.54 L min⁻¹.

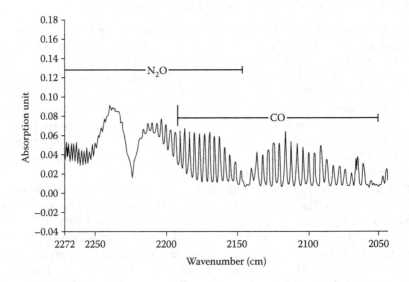

FIGURE 8.10
Zoomed spectrum showing N_2O and CO formation from toluene decomposition at applied voltages of 12.32 kV, discharge gap of 25 mm, and flow rate of 3.54 L min⁻¹.

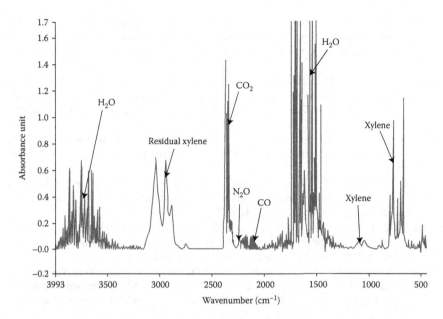

FIGURE 8.11
Spectrum of decomposition of xylene in NTP reactor showing the by-products at 12.32 kV, discharge gap of 25 mm, and flow rate of 3.54 L min^{-1}.

8.9 Conclusion

In this research, treatment of toluene and xylene from wastewater was studied using integrated air stripping and a ferroelectric packed-bed NTP reactor system. The by-product of their decomposition in the NTP reactor was done using FTIR. From the results, 99.7% and 88.47% removal efficiencies were obtained at 50°C and air-water ratios of 60 f toluene and xylene, respectively, using air stripper. The decomposition of toluene and xylene in the NTP reactor increased as the applied voltage increased with 93.27% and 86.89% performance removal efficiency achieved, respectively. Then, the H_2O, CO_2, CO, and N_2O gases that were detected from the FTIR spectrum are the byproducts of toluene and xylene decomposition.

References

Abdullahi, M. E., Abu Hassan, M. A., Zainura, Z. N., and Raja Ibrahim, R. K. 2013a. Simulation of the effect of process variables on packed column air stripper performance. *World Applied Sciences Journal.* 25(7), 1100–1106.

Abdullahi, M. E., Abu Hassan, M. A., Zainura, Z. N., and Raja Ibrahim, R. K. 2013b. Volatile organic compounds abatement from industrial wastewater: Selecting the appropriate technology. *Australian Journal of Basic and Applied Sciences.* 7(12), 103–113.

Anders, S., Teich, T. H., Heinzle, E., and Hungerbhler, K. 1997. VOC treatment with non-thermal plasma. *Proceedings of the 1997 4th International Conference on Advanced Oxidation Technologies for Water and Air Remediation, AOTs-4,* Orlando, FL, USA, 60.

Chen, H. L., Lee, H. M., and Chang, M. B. 2006. Enhancement of energy yield for ozone production via packed-bed reactors. *Ozone Science and Engineering.* 28, 111–118.

Chen, L. H., Lee, M. H., Chen, H. S., and Chuang, B. M. 2008. Review of packed bed plasma reactor for ozone generation and air pollution control. *Industrial and Engineering Chemistry Research.* 47, 2122–2130.

Chuang, K. T., Cheng, S., and Tong, S. 1992. Removal and destruction of benzene, toluene and xylene from wastewater by air stripping and catalytic oxidation. *Industrial Engineering Chemical Research.* 31, 2466–2472.

Chuang, J. S., Kostov, K. G., Urashima, K., Yamamoto, T., Okayasu, Y., Kato, T., Iwaizumi, T., and Yoshimara, K. 2000. Removal of NF3 from semiconductor process flue gases by tanden packed bed plasma and adsorbent hybrid systems. *IEEE Transactions on Industry Application.* 36, 1251–1259.

Creyghton, B. 1997. Direct plasma treatment of polluted water. *Proceedings of the 1997 4th International Conference on Advanced Oxidation Technologies for Water and Air Remediation AOTs-4,* Orlando, FL, USA, 58.

Elbir, T., Cetin, B., Cetin, E., Bayram, A., and Odabasi, M. 2006. Characterization of volatile organic compounds (VOCs) and their sources in the air of Izmir, Turkey. *Environmental Monitoring Assessment.* 133, 149–156.

Fridman, A. 2008. *Plasma Chemistry.* New York, USA: Cambridge University Press.

Goldstein, A. H. and Galbally, I. E. 2007. Known and unexplored organic constituents in the earth's atmosphere. *Journal of Environmental Science and Technology.* 41(5), 1515–1521.

Gomez, E., Amutha Rani, D., Cheesman, C. R., Deegan, D., Wise, M., and Boccaccini, A. R. 2009. Thermal plasma technology for the treatment of wastes: A review. *Journal of Hazardous Materials.* 161, 614–626.

Hammer, T. 1999. Application of plasma technology in environmental techniques. *Contributions to Plasma Physics.* 39(5), 441–462.

Jarrige, J. and Vervisch, P. 2006. Decomposition of three volatile organic compounds by nanosecond pulsed corona discharge: Study of by-product formation and influence of high voltage pulse parameters. *Journal of Applied Physics.* 99, 113303; doi: 10.1063/1.2202700.

Kim, H. H., Prieto, G., Takahima, K., Katsura, S., and Mizuno, A. 2002. Performance evaluation of discharge plasma process for gaseous pollutant removal. *Journal of Electrostatics.* 55, 25–41.

Koutsospyros, A., Yin, S. M., Christodoulatos, C., and Becker, K. 2004. Destruction of hydrocarbons in non-thermal, ambient-pressure, capillary discharge plasmas. *International Journal of Mass Spectrometry.* 233, 305–315.

Li, J., Li, J., Liang, W., Ma, D., Zheng, F., and Jin, Y. Q. 2011. Abatement of toluene from gas streams by using dielectric barrier discharge. *Proceedings of the 2011 IEEE International Conference on Business Management and Electronic Information,* Guangzhou: IEEE, 819–822.

Li, J., Zhu, T., Fan, X., and He, W. 2007. Decomposition of dilute VOCs in air by streamer discharge. *International Journal of Plasma and Environmental Science and Technology.* 1(2), 141–144.

Liang, W., Li, J., Li, J., and Jin, Y. 2009. Abatement of toluene from gas streams via ferroelectric packed bed dielectric barrier discharge plasma. *Journal of Hazardous Materials.* 170, 633–638.

Locke, R. B., Sato, M., Sunka, P., Hoffmann, M. R., and Chang, J. S. 2006. Electrohydraulic discharge and non-thermal plasma for water treatment. *Industrial Engineering Chemical Resources.* 45, 882–905.

Malik, M. A. 2010. Water purification by plasmas: Which reactors are most energy efficient? *Plasma Chemistry Plasma Processing.* 30, 21–31.

Mourad, K., Berndtsson, R., Abu-Elsha'r, W., and Qudah, M. A. 2012. Modelling tool for air stripping and carbon adsorbers to remove trace organic contaminants. *International Journal of Thermal and Environmental Engineering.* 4(1), 99–106.

Navaladian, S., Janet, C. M., Viswanathan, B., and Viswanath, R. P. 2007. On the possible treatment procedure for organic contaminants. *Research Signpost.* 37/661(2), 1–51.

Nehra, V., Kumar, A., and Dwivedi, H. K. 2008. Atmospheric non thermal plasma sources. *International Journal of Engineering.* 2(1), 53–68.

NPI. 2009. *National Pollution Inventory: Volatile Organic Compound Definition and Information.* Australia: Department of Environment, Water, Heritage and the Arts. 1–5.

Raja Ibrahim, R. 2012. Mid-infrared diagnostics of the gas phase in non-thermal application. PhD, University of Manchester, Manchester, London.

Sanchez, M., Karnae, S., and John, K. 2008. Source characterization of volatile organic compounds affecting the air quality in a coastal urban area of south Texas. *International Journal of Environmental Research and Public Health.* 5(3), 130–138.

Schultz, T. E. 2005. Biotreating process waste water: Airing the options. *Chemical Engineering Magazine.* Retrieved November 1, 2011 from http://www.che.com.

Subrahmanyam, C. 2009. Catalytic non-thermal plasma reactor for total oxidation of volatile organic compounds. *Indian Journal of Chemistry.* 48A, 1062–1068.

USEPA. 2011. *National Air Pollution Trends.* Publication of U.S. Environmental Protection Agency. Retrieved September 30, 2012 from http://www.epa.gov/iaq/voc.html.

Vandenbroucke, A. M., Dinh, M. T. N., Giraudon, J., Morent, R., Geyter, N. D., Lamonier, J., and Leys, C. 2011a. Qualitative by-product identification of plasma-assisted TCE abatement by mass spectrometry and FT-IR. *Plasma Chemistry and Plasma Processing.* 31, 707–718.

Vandenbroucke, A. M., Morent, R., Geyter, N. D., and Leys, C. 2011b. Non-thermal plasmas for non-catalytic and catalytic VOC abatement. *Journal of Hazardous Materials.* doi: 10.1016/j.jhazmat.2011.08.060.

Yu-fang, G., Dai-qi, Y., and Ke-fu, C. 2006. Toluene removal characteristics by a superimposed wire plate dielectric barrier discharge plasma reactor. *Journal of Environmental Sciences.* 18(2), 276–280.

Zorgoski, J. S., Carter, J. M., Ivahnenko, T., Laphan, W. W., Moran, M. J., Rowe, B. L., Squillace, P. L., and Toccalino, P. L. 2006. *The Quality of Nation's Water—Volatile Organic Compounds in the Nation's Ground Water and Drinking Water Supply Wells.* USA: U.S. Department of the Interior.

9

Removal of Color Wastewater Using Low-Cost Adsorbent: A Comparative Study

Venmathy Samanaseh, Mohd Ariffin Abu Hassan, and Zainura Zainon Noor

CONTENTS

9.1 Introduction

Comparing various known forms of pollution, water pollution is the greatest concern since water is a prime necessity of humankind and very essential for all living things to survive. However, water pollution has become a universal phenomenon that is more serious in developing countries where it is necessary to consider the discharge of mostly untreated or partially treated municipal and industrial wastewater that harms aquatic habitats (Sharma et al., 2007). Years of increased industrial, agricultural, and domestic activities have resulted in the release of huge amounts of wastewater with a high composition of toxic pollutants.

The availability of clean water is becoming very low and very challenging. Moreover, the demand for water ("Water for People Water for Life" United Nations World Water Development Report UNESCO) has risen for the industrial, agricultural, and domestic sectors which consume 70%, 22%, and 8% of available clean water, respectively. One of the important classes of pollutants is colors. Colors have synthetic origin and are difficult to treat as colors are made of complex molecular structures that make them more rigid, unbreakable, and nonbiodegradable (Forgacs et al., 2004). Colors adsorb sunlight which affects the intensity of light absorbed by hydrophytes and phytoplankton and slows down photosynthesis and dissolved oxygen in the aquatic organisms and it eventually results in high chemical oxygen demand (COD) (Rangabhashiyam et al., 2013). Colored effluent is made of harmful organic and inorganic chemicals that exhibit toxic and carcinogenic effects toward the biological environment. Industries like textile, dyestuffs, paper, and plastics use color for their production and generate colored wastewater.

Mostly, textile wastewater is the major contributor of colored wastewater. Textile wastewater is a mixture of colorants (colors and pigments) and organic compounds used as cleaning agents. The textile industry is one of the largest users of complex chemicals and water during textile processing (Verma et al., 2012). The waste materials from textile wastewater consists of a large portion of biochemical oxygen demand (BOD), COD, pH, temperature, turbidity, toxic chemicals, high concentrations of heavy metal, and total dissolved solids (Sharma et al., 2007). Color is the initial and first contaminant to be found and identified in the wastewater. Comparing different techniques of color removal from wastewater, it is proven that the adsorption technique is one of the best technologies and presented good results in the removal of different types of coloring materials. Many studies have investigated and stated that alternative adsorbents (low-cost adsorbents) are favorable to be used compared to activated carbon due to the restriction of high cost. However, using adsorbents on a large scale is quite expensive for the removal of color. Thus, many researchers are studying the application of low-cost adsorbent as an alternative to costly adsorbents (Ponnusami et al., 2008).

The variety of adsorption in inorganic and organic matrices has been measured and their capacity to remove colors has been studied in a previous paper (Forgacs et al., 2004). Ho and McKay (1998) also stated that adsorption is one of the treatment methods that gives the finest results in removing different coloring materials. This is because adsorption removes the entire color molecule without leaving fragments in the effluent. The extraction method is used to characterize the adsorbent from plant fibers by investigating the origin of the fibers (Sheltami et al., 2012). The effectiveness of this extracted cellulose is investigated for the first time in treating methylene blue. The adsorption is defined as percentage reduction of initial color concentration and COD level.

9.2 Color Removal Technologies in Wastewater

Assorted treatment methods are available to remove color in wastewater. In general, there are three main classifications of wastewater treatment methods. These consist of chemical, physical, and biological treatments. The method of removing color of wastewater is one effective way to control water pollution and standard water quality from hazard. The color removing wastewater treatment has been earlier stated by Pokhrel and Viraraghavan (2004), Robinson et al. (2001), and Banat et al. (1996). Currently, due to the complicated nature of wastewater, a single process is capable of inadequate treatment. Thus, many studies are in progress to find joint ventures of various methods to achieve favorable water quality in an economical way. These technologies in removing color wastewater industrial can be divided into three categories as stated above. These processes have several variable techniques effective for certain types of wastewater and certain coloring substances.

9.3 Adsorption

Table 9.1 shows adsorption is more effective than other color removal technologies. Adsorption technique is the physical treatment of wastewater that economically meets higher effluent standards and water reuse requirement. It is the most effective technique and feasible without having any result in the formation of harmful substances. The process of adsorption is sludge-free clean operation and colors could be fully removed, even from the diluted solution. Adsorption is also well known as an equilibrium separation process and it is an efficient process for water purification application. It has been found to be superior to other techniques for water reuse in terms of flexibility and simplicity of design, initial cost, ease of operation, and insensitivity to toxic pollutants. Previous research has been done using coconut shell carbon, activated carbon, activated alumina, and many others as adsorbents.

Adsorption has the potential to substitute the processes of distillation, absorption, and liquid extraction to remove trace components from the liquid phase. Separation by adsorption relies on the capability of the component of being adsorbed more than others. Three different modes of contacting the adsorbent and wastewater are batch contact, fixed-bed contact, and fluidized-bed contact. Among the three of them, the batch contact system is highly effective on a smaller scale of operation. While that, the fixed-bed contact is more effective on constant dye concentration with the adsorbent at all times and the fluidized-bed system consists of high rate of mass transfer where the operation is high critical in flow rates and loading volume.

There are three steps in adsorption. Figure 9.1 describes the overall diffusion and adsorption process. Firstly, the adsorbate diffuses from the major stream to the external surface of the adsorbent particle. Secondly, the adsorbate moves from the smallest area of the external surface to the pores within each adsorbent particle. A large amount of adsorption normally happens in these pores

TABLE 9.1

Comparison of Color Removal Technologies in Industrial Wastewater

Color Removal Technology	Type of Wastewater	Application	Pros	Cons	% of Color Removal	Reference
Microbial degradation	–	Variety of microorganisms and approaches that result in effective systems. Mostly fungal decolorization, adsorption by (living or dead) microbial biomass and bioremediation systems are applied to wastewater treatment	Ensures the recuperation of valuable components from effluent streams. It has short-term utilization due to the wide variety of membrane shapes, modules, and materials commercially available. Environmental friendly bacteria, yeasts, algae, and fungi are capable of digesting different pollutants	Short operational stability of the process, low specific capacity, less possibilities for the retention of the enzyme and its application for detoxification of other recalcitrant compounds	–	López et al. (2002)
Advanced oxidation process (AOP)	Pulp mill effluent	Effectively changes the properties of a carbonaceous material specifically its affinity toward polar adsorbate. Oxidizing agents oxidize recalcitrant organic pollutants to stable inorganic (H_2O_2 CO_2) or biodegradable compounds	Potentially powerful method and capable of changing the pollutants into harmless substances	Requires more reaction time and larger reactor volumes. Sludge generation and its handling	80%–91%	Shahbazi et al. (2014) and Catalkaya and Kargi (2007)

(Continued)

TABLE 9.1 (Continued)

Comparison of Color Removal Technologies in Industrial Wastewater

Color Removal Technology	Type of Wastewater	Application	Pros	% of Color Removal	Cons	Reference
Photo catalytic degradation	Textile wastewater	Uses photoactive catalysts illuminated with UV light to produce highly reactive radicals that can convert the organic compounds to harmless substances like CO_2 and water	Spent adsorbents could be successfully regenerated using this process. It does not add any further to pollution. The active oxidizing species, the hydroxyl radicals, will dimerize to give hydrogen peroxide, which may degrade ultimately to water and oxygen	—	Slow electron transfer	Crittenden et al. (1997) and Sun et al. (2008)
Coagulation and flocculation	Textile wastewater	More accomplished method which can be normally used with biological treatment, to remove suspended solids and organic matter (colors)	Fast detention time and easy to operate	78%–93%	Type of coagulants or flocculants and pH	Tünay et al. (1996) and Verma et al. (2012)
High-rate clarification	Industrial wastewater	Operates chemical and physical treatment and uses flocculation and sedimentation processes to reach rapid settings of color	The units are compact and the space requirements have been reduced. The start-times are quick (less than 30 min) to reach efficiency and highly clarified effluent is emitted	—	—	Metcalf and Eddy (2003)
Adsorption	Textile wastewater	A process of accumulating substances which present in solution on a suitable interface	Adsorption process was more than 96% efficient for treatment of water. Most effective technique and feasible without have any result in the formation of harmful substances	97%	Usage of adsorbents is very expensive	Crittenden et al. (1997), Robinson et al. (2001), Metcalf and Eddy (2003), and Garg et al. (2003)

Contaminant molecules

 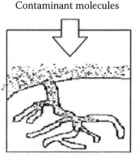 

Step 1: Diffusion to   Step 2: Migration into   Step 3: Layer buildup
adsorbent surface   pores of adsorbtion   of adsorbate

FIGURE 9.1
Process of adsorption.

because there is abundant available surface on particles. At last, the polluted particles are attached to the surface in the pores (Walther et al., 1988).

According to Metcalf and Eddy (2003), adsorption process contains three steps which are macrotransport, microtransport, and sorption. Macrotransport is where the adsorbate moves through water to the liquid/ solid interface by the process of advection and diffusion. Microtransport is where the adsorbate diffuses through the macropore system of the solid adsorbent to the adsorption sites in the micropores. Mostly adsorption is carried out on the surface of the solid adsorbent and in the macropores and mesopores. The surface area of those parts is very tiny compared to the micropores. Thus, the content adsorbed by micropores is usually negligible.

Current research (Table 9.2) has shown that adsorption has been used successfully and favorably in color removal from industrial effluents as compared in Table 9.1. Mostly, activated carbon is used as adsorbent. However, the search for alternative adsorbents is highly anticipated as its cost is high and an enormous quantity of effluents is produced by industries daily (Low et al., 2000). Low-cost adsorbents are the secondary choice for adsorption for color emitting industries like plastic, paper, textile, and cosmetics which use color in their processes that leads to water pollution.

9.4 Color

Colors are the primary pollutant in wastewater (Banat et al., 1996). It is considered hazardous to the environment because they affect the nature of water and reduce photosynthetic action. However, color is an important substance in the textile industry. Ninety percent of total colors consumption

TABLE 9.2

Adsorption Technologies to Remove Color of Industrial Wastewater

Comments	Adsorbent	Type of Color Removed	Reference
Adsorption is highly efficient for the removal of color in terms of initial cost, simplicity of design, ease of operation, and insensitivity to toxic substances. The study done to determine the optimum conditions for the removal of a dye, methylene blue, from simulated wastewater by formaldehyde treated sawdust and sulfuric acid treated sawdust carbon of Indian Rosewood	Indian Rosewood (*Dalbergia sissoo*) sawdust	MB	Garg et al. (2003)
In this work, titanium peroxide powder is successfully prepared and the material exhibits excellent selective adsorption performance for cationic dyes. The material was characterized by several techniques. Adsorption technique made the decolorization process very fast	Titanium peroxide powder	MB, MG, and NR	Zhao et al. (2014)
Adsorption was recognized to be a promising and a cost-effective process to remove colors from aqueous solution. It is a rapid process, spontaneous, and endothermic	Spent activated clay	MB	Weng and Pan (2007)
Adsorption is dependent on the oxygenated groups at the surface of adsorbent	Globe artichoke leaves	MB	Benadjemia et al. (2011)
Adsorption helps to give high percentage of color removal	Almond shells	Direct Red 80 (DR 80)	Ardejani et al. (2008)

ends up on fabrics and it is also estimated that 280,000 tonnes of textile colors are discharged into industrial wastewater globally (Hameed et al., 2007). The global demand for colors and pigments is expected to increase rate of 3.5% annually, from 1.9 million tonnes in 2008 to 2.3 million tonnes in 2013. Synthetic colors are mostly used in many fields with various technologies, such as the textile industry, paper production, light harvesting arrays, agricultural research, and the leather tanning industry, food technology, photoelectrochemical cells, and hair coloring.

These organic compounds are normally water-soluble or water dispersible which is capable of being adsorbed into the substrate to devoid the crystal structure of the substance. The molecules are chemically bonded to the substance and become part of it. Most colors are complex organic molecules and resistant to weather and the action of detergents. The color intensity of color molecules is between 400 and 700 nm which depends on their ability of adsorption radiation in the visible region. Characterization of color is based on their chemical structure or by the method it is applied to the substrate. Moreover, colors can be classified by their chemical type and the application. Color is usually expressed in terms of a color index (CI).

Chromophore is a group of atoms that is responsible for color, as well as an electron withdrawing or donating substituent that intensify the color of the chromophores, called auxochromes (Christie, 2001). According to Lucas and Peres (2006), there are 20–30 different types of colors that can be detected based on chromophores. The most important chromophores are azo (–N=N–), carbonyl (–C=O), methine (–CH=), nitro (–NO$_2$), and quinoid groups. The most important auxochromes are amine (–NH$_3$), carboxyl (–COOH), sulfonate (–SO$_3$H), and hydroxyl (–OH). Textile colors are divided into three main groups which are anionic (direct, aid, and reactive colors), cationic (basic colors) and nonionic (disperse colors) (Robinson et al., 2001). All the colors can be classified through their chromophores and group. Cationic colors are also known as basic colors which have high bright concentration of colors and are highly noticeable even in very low intensity (Banat et al., 1996). As cationic colors are toxic, carcinogenic, mutagenesis, chromosomal fractures, teratogenicity, and respiratory toxicity, it has to be eliminated from wastewater.

Cationic colors have various applications such as printing, paper, textiles, electroplating, food, cosmetics, and pharmaceutical products that cause them to be the main contributors to water pollution (Zhao et al., 2014). Previous studies stated that the mutagenic activity of textile wastewater effluents, using the salmonella/microsome assay contributed the highest percentage (67%) of mutagenic effluents. The microsome assay is a test to detect a broad range of chemical substances that can generate genetic damage that brings gene mutations. Methylene blue (MB), malachite green (MG), and crystal violet (CV) are popular cationic colors used widely in various fields including textiles, rubber, paper, and plastics (Saitoh et al., 2014).

Chromophore is a group of atoms responsible for color. MB is a water-soluble cationic color when referring to their chromophore (Ahmad et al., 2007). The molecular formula of MB is C$_{16}$H$_{18}$N$_3$SCl and it is a heterocyclic aromatic chemical compound. Figure 9.2 shows the chemical structure of MB. At room temperature MB is a solid state, odorless, dark green powder that yields a blue solution when dissolved in water. The hydrated molecule of MB forms three molecules of water.

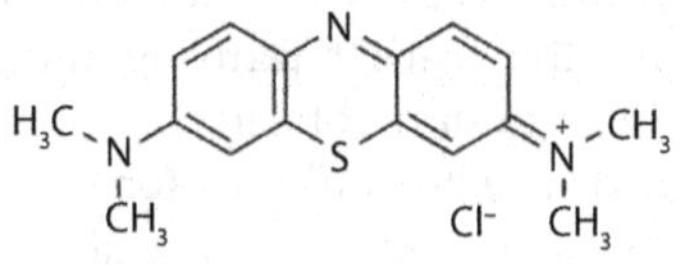

FIGURE 9.2
Chemical structure of methylene blue. (Adapted from Weng, C. H. and Pan, Y. F. 2007. *Journal of Hazardous Materials*, 144, 355–362.)

9.5 Low-Cost Adsorbents

A wide variety of adsorbents have been extensively used in effluent treatment. Cost-effective adsorbents for treatment of contaminated water streams are needed. Natural materials that are abundant in large quantity, or industrial and agricultural wastes are the target to be transformed into inexpensive adsorbents. Cost is a crucial matter to be compared when choosing the adsorbents. Low-cost adsorbents cause the adsorbents to be expended where it can be disposed without costly regeneration. Adsorbent is classified as "low cost" if it needs low processing effort, is abundant in nature, or if it is wasted material or a by-product from other industries (Bailey et al., 1999). Improved technology may compensate for the cost of additional processing. Some research has been conducted on a wide range of low-cost adsorbents. Solid adsorbents should have two main criteria which are large surface area to volume ratio and preferred affinity for certain components in the liquid phase.

Table 9.3 shows some leaves used as low-cost adsorbents to remove color in industrial wastewater. Leaves listed in the table have similar abundance as *Melastoma malabathricum* leaves discussed in this paper. Thus, the performance in color removal of the selected leaves is used as reference for *Melastoma malabathricum* leaves.

9.5.1 *Melastoma malabathricum* Cellulose

Melastoma malabathricum is a flowering plant in the family of Melastomataceae. It is locally called as Senduduk, Hosing, Karamunting, Kemunting, Kudok-kudok, Odok-odok, and Seppabang in Malaysia. It is a very common herb that is abundant in moist places of tropical areas of Asia, such as India, Thailand, and Malaysia, and the Pacific. It normally grows as a small tree for about 12–13 ft or even up to 20 ft. This species has three types of flowers, namely large, medium, and small-sized flowers with dark purple-magenta petals, light pink-magenta petals, and white petals. The following is the scientific classification of the leaves:

Kingdom:	Plantae
Unranked:	Angiosperms
	Eudicots
	Rosids
Order:	Myrtales
Family:	Melastomataceae
Genus:	Melastoma
Species:	*M. malabathricum*

TABLE 9.3

Leaves as Low Adsorbents to Remove Color in Industrial Wastewater

Leaves	Color Removed	Description	Results					Reference
			Adsorption Capacity	Adsorbent Dosage	Time of Agitation	pH		
Neem (*Azadirachta indica*)	Green	Neem leaves belong to the Meliaceae family, from the Indian subcontinent. The seeds and leaves from this tree have been applied since early days for the treatment of human ailments and for medication. The leaves washed repeatedly with water to get rid of dust and water soluble impurities, dried, ground, and sieved. Color concentration, adsorbent dosage, and pH were investigated in batch adsorption	133.69 mg/g	2 g/L	5 h	2–8		Bhattacharyya and Sharma (2003), Bhattacharyya and Sharma (2004), and Bhattacharyya and Sharma (2005)
Guava (*Psidium guajava*)	Methylene blue	The leaves from the Myrtaceae family, a tropical and semitropical plant. As neem leaves, guava leaves and seeds have medicinal value and have traditionally been used in the treatment of human ailments. As they are abundantly available and low cost they can be disposed of after use without a further expensive method of regeneration. The leaves were washed repeatedly with water to get rid of dust and water soluble impurities, dried, ground, and sieved. Then, the adsorbent of guava leaves was applied to remove color and the result proved that, it gave better color removal	133.33 mg/g	500 mg/dm³	120 min	7.5		Ponnusami et al. (2008)

(Continued)

TABLE 9.3 (*Continued*)

Leaves as Low Adsorbents to Remove Color in Industrial Wastewater

Leaves	Color Removed	Description	Results					Reference
			Adsorption Capacity	Adsorbent Dosage	Time of Agitation	pH		
Teak leaves		It was found that teak leaf powder possesses good adsorption potential and removes color from aqueous solutions effectively. The thermodynamic parameters used in the study concluded that the adsorption of methylene blue by teak leaf powder is continuous, chemisorptive and exothermic in nature. Other studies also revealed that teak leaf powder has excellent adsorption capacity of textile industry effluent	219 mg/g	2.0 mg/dm^3	–	7		Ponnusami and Srivastava (2009)
Acacia nilotica leaves	Crystal violet (CV) and rhodamine B (RHB)	*Acacia nilotica* leaves originated from Africa and the Indian subcontinent. The effect of solution pH, agitation time, and initial color concentration were evaluated. *Acacia nilotica* leaves act as ecofriendly agricultural waste for the removal of colors (CV and RHB) in the textile industry. CV and RHB are toxic to some organisms causing direct destruction of aquatic communities. They can cause allergic dermatitis, skin irritation, cancer, and mutation in human beings	33 mg/g for CV and 37 mg/g for RHB	0.2–1 g	180 min	6		Prasad and Santhi (2012)

FIGURE 9.3
Melastoma malabathricum leaves.

Melastoma malabathricum is a green foliage plant which has simple, ladder-like veins, densely hairy and whitish-yellow colored lower surface of leaves as shown in Figure 9.3.

The flowers are averagely sized about 35 mm diameter and placed in feathered clusters. The fruits are averagely sized about 9 mm in diameter, yellowish-gray and fleshy capsules with hairs. Unfortunately, the chemistry of Melastomataceae is lacking. It is mostly used for medical purposes among the Malay community. It acts as medicine for diarrhea, puerperal infection, dysentery, leucorrhea, wound healing, postpartum treatment, and hemorrhoids. The roots can be used as a mouthwash for toothache. The fruits can also be eaten. It is also used for phyto remediation as stated in a few previous studies.

Cellulose was firstly discovered around year 1838 by Anselme Payen in France. According to Liu and Sun (2010), cellulose is the most abundant polymer available in nature with the estimation of natural production of 1.5×10^{12} tonnes. It is considered as the primary source of raw materials. It can be found on the walls of plant cells and several marine animals. Cellulose is the main component of the cell wall in lignocellulosic plants, 23%–53% on dry-weight basis. The content of cellulose varies from each plant according to maturity, position, growing environment, growth, and species.

Plant cell wall is the most supportive for plants and limits the shape and size of cells and acts as barrier to potential pathogens. In most straw walls, cellulose lies as microfibrils of indefinite length and it causes the crystallinity to vary. This is because microfibrils contribute mechanical strength to the cell wall and act as a framework to the cell wall. Cellulose exists as a gel matrix consisting of hemicelluloses, lignins, and other carbohydrate polymers. Cellulose is a liner homopolymer composed of d-glucopyranose units linked by β-1,4-glycosidic bonds ($C_6H_{10n+2}O_{5n+1}$ (n = degree of polymerization of glucose)) as shown in Figure 9.4. Cellulose consists roughly of 6000 glucose units.

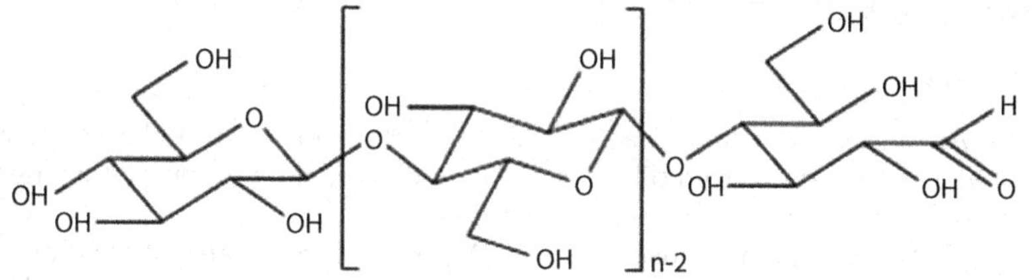

FIGURE 9.4
Chemical structure of cellulose.

The cellulose of a plant can be extracted by chemical treatment methods. Out of many effects made to isolate cellulose from various biomass sources, alkali extraction facilitates most effectively for separating cellulose from straws. Alkali treatment involves treatment with sodium hydroxide (NaOH) and bleaching with sodium chlorite ($NaCIO_2$). Previous studies show that in the usage of chemical treatment in cellulose extraction from empty fruit bunches, the cellulose content increased from 40% to 90% and the amount of lignin and hemicelluloses decreased from 4% to 2%. NaOH acts as swelling agent to increase the accessibility of core material for the bleaching process.

Cellulose can be partially degraded by this alkali treatment. As the rate of cellulose degradation was reduced through this method, the cellulose chains would be less damaged. Bleaching is to make sure the most of lignin is removed from the fibers. The temperature of decomposition for cellulose, hemicelluloses, and lignin are different as their structures vary. Thus, the fibers and the cellulose will have a small weight loss in the temperature range of 25–150°C. Cellulose degrades if the temperature is higher than 250°C after alkali treatment and bleaching. Through cellulose extraction the amorphous region of cellulose is removed and crystallinity and thermal stability are improved.

9.6 Conclusion and Suggestion

Adsorption is justified as the appropriate method to remove color as it has more advantages. Low-cost adsorbents are suggested to substitute adsorbents (active carbon) that are costly in adsorption process. *Melastoma malabathricum* cellulose is discussed in this chapter to remove methylene blue color from wastewater as it is abundant locally in Malaysia. Methylene blue is one of the cationic colors that are mostly emitted from wastewater. This review shows that *Melastoma malabathricum* can be used as a low-cost, environmentally friendly substitute adsorbent for wastewater treatment.

References

Ahmad, A. L., Loh, M. M., and Aziz, J. A. 2007. Preparation and characterization of activated carbon from oil palm wood and its evaluation on methylene blue adsorption. *Dyes and Pigments*, 75, 263–272.

Ardejani, F. D., Badii, Kh., Yousefi Limaee, N., Shafaei, S. Z., and Mirhabibi, A. R. 2008. Adsorption of Direct Red 80 dye from aqueous solution onto almond shells: Effect of pH, initial concentration and shell type. *Journal of Hazardous Materials*, 151, 730–737.

Bailey, S. E., Olin, T. J., Bricka, R. M., and Adrian, D. D. 1999. A review of potentially low-cost sorbents for heavy metals. *Water Research*, 33(11), 2469–2479.

Banat, I. M., Nigam, P., Singh, D., and Marchant, R. 1996. Microbial decolourisation of textile-dye-containing effluents: A review. *Bioresource Technology*, 58, 217–227.

Benadjemia, M, Millière, L., Reinert, L., Benderdouche, N., and Duclaux, L. 2011. Preparation, characterization and methylene blue adsorption of phosphoric acid activated carbons from globe artichoke leaves. *Fuel Processing Technology*, 92, 1203–1212.

Bhattacharyya, K.G. and Sarma, A. 2003. Adsorption characteristics of the dye, brilliant green on neem leaf powder. *Dyes and Pigments*, 57, 211–222.

Bhattacharyya, K. G. and Sharma, A. 2004. Adsorption of Pb(II) from aqueous solution by *Azadirachta indica* (neem) leaf powder. *Journal of Hazardous Materials*, B11, 97–109.

Bhattacharyya, K. G. and Sharma, A. 2005. Kinetics and thermodynamics of methylene blue adsorption on neem (*Azadirachta indica*) leaf powder. *Dyes and Pigments*, 65, 51–59.

Catalkaya, E. C. and Kargi, F. 2007. Color, TOC and AOX removals from pulp mill effluent by advanced oxidation processes: A comparative study. *Journal of Hazardous Materials*, B139, 244–253.

Christie, R. 2001. *Colour Chemistry*. The Royal Society of Chemistry, Cambridge, United Kingdom.

Crittenden, J. C., Suri, R. P. S., Perram, D. L., and Hand, D. W. 1997. Decontamination of water using adsorption and photocatalysis. *Water Research*, 31(3), 411–418.

Forgacs, E., Cserháti, T., and Oros, G. 2004. Removal of synthetic dyes from wastewaters: A review. *Environment International*, 30(7), 953–971.

Garg V. K., Amita, M., Kumar, R., and Gupta, R. 2003. Basic dye (methylene blue) removal from simulated wastewater by adsorption using Indian Rosewood sawdust: A timber industry waster. *Dyes and Pigments*, 63, 243–250.

Hameed, B. H., Ahmad, A. A., and Aziz, N. 2007. Isotherms, kinetics and thermodynamics of acid dye adsorption on activated palm ash. *Chemical Engineering Journal*, 133(1–3), 195–203.

Ho, Y. S. and Mckay, G. 1998. Sorption of dye from aqueous solution by peat. *Chemical Engineering Journal*, 70(2), 115–124.

Liu, C. F. and Sun, R. C. 2010. Cereal Straw as a Resource for Sustainable Biomaterials and Biofuels: Chapter 5—Cellulose *Chemistry, Extractives, Lignins, Hemicelluloses and Cellulose*, pp. 131–167.

López, C., Mielgo, I., Moreira, M. T., Feijoo, G., and Lema, J. M. 2002. Enzymatic membrane reactors for biodegradation of recalcitrant compounds. Application to dye decolourisation. *Journal of Biotechnology*, 99(3), 249–257.

Low, K. S., Lee, C. K., and Liew, S. C. 2000. Sorption of cadmium and lead from aqueous solutions by spent grain. *Process Biochemistry*, 36, 59–64.

Lucas, M. S. and Peres, J. A. 2006. Decolorization of the azo dye reactive black 5 by fenton and photo-fenton oxidaiton. *Dyes and Pigments*, 71, 236–244.

Metcalf and Eddy. 2003. *Wastewater Engineering: Treatment and Reuse*, 4th ed., Mcgraw-Hill, New York, pp. 293–295, 411–412, 1138.

Pokhrel, D. and Viraraghavan, T. 2004. Treatment of pulp and paper mill wastewater—A review. *Science of the Total Environment*, 333(1–3), 37–58.

Ponnusami, V. and Srivastava, S. N. 2009. Studies on application of teak leaf powders for the removal of color from synthetic and industrial effluents. *Journal of Hazardous Materials*, 169, 1159–1162.

Ponnusami, V., Vikram, S., and Srivastava, S. N. 2008. Guava (*Psidium guajava*) leaf powder: Novel adsorbent for removal of methylene blue from aqueous solutions. *Journal of Hazardous Materials*, 152(1), 276–286.

Prasad, A. L. and Santhi, T. 2012. Adsorption of hazardous cationic dyes from aqueous solution onto *Acacia nilotica* leaves as an eco friendly adsorbent. *Sustainable Environmental Research*, 22(2), 113–122.

Rangabhashiyam, S., Anu, N., and Selvaraju, N. 2013. Sequestration of dye from textile industry wastewater using agricultural waste products as adsorbents. *Journal of Environmental Chemical Engineering*, 1(4), 629–641.

Robinson, B. T., Mcmullan, G., Marchant, R., and Nigam, P. 2001. Remediation of dyes in textile effluent: A critical review on current treatment technologies with a proposed alternative. *Bioresource Technology*, 77(3), 247–255.

Saitoh, T., Saitoh, M., Hattori, C., and Hiraide, M. 2014. Rapid removal of cationic dyes from water by coprecipitation with aluminum hydroxide and sodium dodecyl sulfate. *Journal of Environmental Chemical Engineering*, 2, 752–758.

Shahbazi, A., Gonzalez-Olmos, R., Kopinke, F. D., Zarabadi-Poor, P. P., and Georgi, A. 2014. Natural and synthetic zeolites in adsorption/oxidation processes to remove surfactant molecules from water. *Separation and Purification Technology*, 127, 1–9.

Sharma, K. P., Sharma, S., Sharma, S., Singh, P. K., Kumar, S., Grover, R., and Sharma, P. K. 2007. A comparative study on characterization of textile wastewaters (untreated and treated) toxicity by chemical and biological tests. *Chemosphere*, 69(1), 48–54.

Sheltami, R. M., Abdullah, I., Ahmad, I., Dufresne, A., and Kargarzadeh, H. 2012. Extraction of cellulose nanocrystals from mengkuang leaves (*Pandanus tectorius*). *Carbohydrate Polymers*, 88(2), 772–779.

Sun, J., Qiao, L., Sun, S., and Wang, G. 2008. Photocatalytic degradation of orange G on nitrogen-doped TiO_2 catalysts under visible light and sunlight irradiation. *Journal of Hazardous Materials*, 155, 312–319.

Tünay, O., Kabdasli, I., Eremektar, G., and Orhon, D. 1996. Color removal from textile wastewaters. *Water Science and Technology*, 34(11), 9–16.

Verma, A. K., Dash, R. R., and Bhunia, P. 2012. A review on chemical coagulation/flocculation technologies for removal of colour from textile wastewaters. *Journal of Environmental Management*, 93(1), 154–168.

Walther, H. J., Faust, S. D., and Aly, O. M. 1988. Adsorption processes for water treatment. *Acta Hydrochimica et Hydrobiologica*, 16(6), 572.

Weng, C. H. and Pan, Y. F. 2007. Adsorption of a cationic dye (methylene blue) onto spent activated clay. *Journal of Hazardous Materials*, 144, 355–362.

Zhao, X. G., Huang, J. G., Wang, B., Bi, Q., Dong, L. L., and Liu, X. J. 2014. Preparation of titanium peroxide and its selective adsorption property on cationic dyes. *Applied Surface Science*, 292, 576–582.

10

Bioparticle Development in Constructed Wetland for Domestic Wastewater

Mohd. Fadhil Md Din, Zaharah Ibrahim, Zaiton Abd Majid, Chi Kim Lim, and Abdul Hadi Abdullah

CONTENTS

10.1 Domestic Wastewater

Domestic wastewater is typically mixed sewage from residential or commercial areas from sinks, bathtubs, toilets, washing machines, and dishwashers. Furthermore, it is commonly called black water (from toilets) or grey water (from kitchen and bathrooms). The wastewater produced nowadays is getting polluted due to modernization and population growth. Domestic wastewater has an impact on environmental conditions in rivers and coastal waters, especially if improper treatment or insufficient treatment capability is applied. Previously, septic conditions have been the predominant treatment to cater to individual systems which were always hampered by their inefficiency. The impact of the discharge of wastewater includes the unsightly littering of rivers, and the creating of foul smells and potential health hazards. Continued pollution may threaten the survival of aquatic life in rivers. There are more chemicals used and discharged into the sewerage treatment system from kitchens and bathrooms.

10.2 Bioparticle

The bioparticle is initially developed as a hollow particle (composite materials) produced by a mixture of zeolite, calcium hydroxide ($Ca(OH)_2$), and activated carbon that has the potential to treat wastewater, usually the recalcitrant compounds. In the development of the bioparticle, different ratios of the components making up the bioparticle can be used depending on the type of wastewater to be treated. Due to the components (zeolite and activated carbon) which make up the bioparticle, the bioparticle can function as an adsorbent. In recent years, the application of natural and modified zeolite as an ion exchanger has been one of the most effective technologies used to remove various contaminants due to their high ion-exchange capacity, high specific surface areas, and relatively low cost (Crini, 2006). The application of natural zeolites for wastewater treatment systems has been widely used and is still a promising technique in environmental cleaning processes. In the past decades, the application of natural zeolites is more focused on ammonium and heavy metal removal due to the nature of ion exchange. Cations, anions, and organic compounds widely exist in wastewater. In recent years, natural zeolite and its modified forms have also been proven to be effective in the removal of anions and organic compounds from water systems (Wang and Peng, 2010), whereas activated carbon is an amorphous form of carbon which has a high degree of porosity and an extensive surface area (Faria et al., 2004). It has been widely used for the separation of gases, recovery of solvents, removal of pollutants from drinking water, and as a catalyst support. The starting materials usually used for synthesizing activated carbon are coals and ligno-cellulosic materials (Hayashi et al., 2000). Although activated carbons are widely used as an adsorbent in wastewater treatment, the specific mechanisms for the adsorption of activated carbon for organic and inorganic solutes are still unclear. It has been proposed that organic compounds adsorb through a π–π dispersion interaction mechanism which involves the interaction of the π electrons of the aromatic species with the π electrons of the basal planes of the carbon (Chingomb et al., 2005). With a few modifications, the bioparticle will simply be suitable to adapt to many beneficial bacteria and thus, their shape, size, and materials can

TABLE 10.1

Research on Bioparticle Material

Researcher	Material	Ratio
	Zeolite:slaked lime:water	2:1:1
Nurfarahain (2008)	Zeolite:blast furnace slag:water	2:1:2, 2:2:1, 2:1:1
	Zeolite:slaked lime:light weight aggregates:activated carbon:water	3:2:2:1:1
	Zeolite:steel slag:cement:water	4:3:2:1
	Zeolite:steel slag:cement:water	4:3:2:0.5

be drastically changed. The changes are believed to increase the efficiency in treating polluted wastewater compared to the first version of bioparticle development. Bioparticles can be similar to the trickling filter technology which enables those composites to treat a low strength of wastewater at high micronutrient contents. Bioparticles will create a favorable condition for indigenous/bioaugmentated microbes that are immobilized on the surface of the composites to enhance the remedy process. A research study using *Enterococcus faecalis* ZL immobilized onto a macrocomposite (bioparticle) reported that almost complete decolorization of AO7 within 3 h of treatment time and more than 80% of COD removal were achieved by the applied biofilm system. Another study carried out by Lazim et al. (2014) involved the use of biofilm-coated macrocomposites for the treatment of palm oil mill effluent (POME) obtained from the palm oil industry. The biofilm that was coated on the macrocomposites consisted of a mixed bacterial culture of *Brevibacillus panacihumi*, *Enterococcus faecalis*, *Lysinibacillus fusiformis*, and the newly identified *Klebsiella pneumonia* (MABZ). All of these bacteria were previously confirmed as bacteria capable of removing color and COD. During the treatment, results showed color reduction of POME up to 92% (initial 3514 ADMI) and COD removal of 93% (initial 888 mg/L) after 6 days of incubation. The color removal and COD reduction of the final POME were enhanced by biofilm-coated macrocomposites.

In this study, the bacteria that will be used are from previous research stock culture which have been obtained from different places and different environments. These bacteria have proven their effectiveness in treating wastewater. The bacteria will be mixed with the bioparticles in the form of mixed culture (Haidiahvita, 2005). Table 10.1 shows the recent research using various mixtures of composites to produce bioparticles.

10.3 Constructed Wetlands

Constructed wetlands are preferred in land-based treatment applications because the systems are cost effective and require no energy or less energy, and they are easy to control (Ayaz and Akca, 2001). Constructed wetlands have proved to be a very effective method for the treatment of municipal wastewater. As they are depleted and affected by development, the importance of natural wetlands in watershed systems becomes more significant. Efforts to restore and maintain wetlands have been crucial to water quality in many areas.

The use of constructed wetlands to treat urban surface runoff and remove nutrients from diverse sources in rural catchment has received much attention lately. The application of constructed wetlands for municipal wastewater treatment has led to the study of their use for other kinds of wastewater. Therefore, a better understanding in nitrogen transformation will be beneficial for complete treatment in an actual wetland system.

The real drawback of the actual application of wetlands is their land acquisitions. The integration of bioparticles is then efficient to reduce the soil area without influencing the bioremediation activity. It is reported that any circumstances of flood, storm water, and any other natural disaster will significantly reduce the pollutant removal efficiencies. Therefore, bioparticle combination with constructed wetlands will be useful to apply without delay problems related to uncertainty.

10.4 Scope of Study

The experiments focused on the wastewater characteristics, removal of pollutants, development of the bioparticle, and construction in lab-scale subsurface wetlands carried out in the Environmental Laboratory, Faculty of Civil Engineering, Universiti Teknologi Malaysia. The lab-scale system of integrated bioparticle and subsurface wetlands has been fabricated into four compartments. The domestic wastewater sample was taken from an individual septic tank (IST) or Imhoff tank (IT) at Taman UngkuTunAminah (TUTA), Johor Bahru. The vegetation species for wetland plant is *heliconia*, a species from the *Musaceae* family. A small development of cubical shape of bioparticle has been used in the study; made from zeolite, slake lime, and blast furnace slag immobilized with indigenous microbes. The combination of wetlands and bioparticle is an alternative to reuse the land application. The suggested parameters to be monitored during the experimental operations are

1. Biological oxygen demand (BOD)
2. Chemical oxygen demand (COD)
3. Ammonia nitrogen (AN)
4. Phosphorus (P)
5. Nitrate (NO_3)

10.5 Environmental Issues from Septic Tanks

Although a septic tank is a popular on-site sewage treatment system, it still exhibits some potential problems which raise environmental issues that need to be acknowledged. The main problem in the septic tank is oil and grease. An excessive dumping of cooking oils and grease can fill up the upper portion of the septic tank and can cause the inlet drains to block. Oils and grease are often difficult to degrade and can cause odor problems and difficulties with the periodic emptying.

Another issue is nutrient contamination such as of nitrogen and phosphorus in septic tank effluent. Nitrogen principally occurs in organic and ammoniacal forms in sewage and in septic tank effluent, while phosphate is a common component of domestic waste and currently considered to be a major cause of eutrophication. Septic tanks are ineffective in removing nutrient compounds that can cause algae blooms in receiving waters and lead to human health hazards.

In areas with high population density, groundwater pollution levels often exceed acceptable limits. Some small towns are facing the costs of building very expensive centralized wastewater treatment systems because of this problem, owing to the high cost of extended collection systems. Ensuring existing septic tanks are functioning properly can also be helpful for a limited time, but becomes less effective as a primary remediation strategy as population density increases. Hence, septic tanks need additional treatment for better septic operation and wastewater treatment in order to get a cleaner effluent to discharge into groundwater or water sources.

10.6 Research Methodology

This study initially started with the fabrication of wetlands lab-scale system, which also consisted of a storage tank of 500 L and four cells of wetlands with an effluent collection tank. All laboratory equipment mentioned were made from fabric plastic, perspex, and PVC. The cell is fabricated in rectangular shape with dimensions of 0.3 m width, 0.3 m length, and 0.5 m depth. As shown in Figure 10.1, the lab-scale system of bioparticle and subsurface wetlands was constructed into four sets of configurations determined as control, wetlands, bioparticle, and combination of bioparticle in wetlands. The first compartment in the wetland cell is planted with *Heliconia rostrata*, while second compartment is filled with bioparticle without plants. The third compartment is planted with *Heliconia rostrata* and bioparticle, and last compartment is without plants (control).

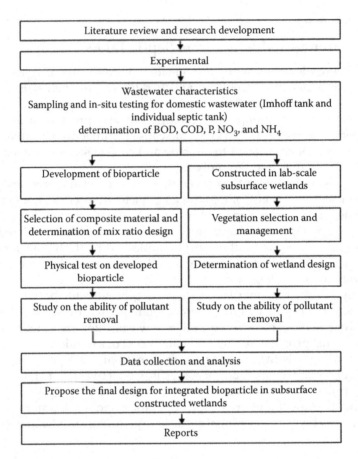

FIGURE 10.1
Research framework and development.

Next, the development of the bioparticle is determined by selection of composite material and physical testing. This involves mixing with other materials and an appropriate volume of water. The bioparticle was developed using zeolite, slake lime, and blast furnace slag with a minimal volume of water. Blast furnace slag and zeolite were crushed together using a grinder and then the particle was sieved through 0.18 mm. The materials were mixed together based on a suitable mix ratio design. After the mixing with ratio 4:3:2:1 (zeolite:blast furnace slage:slake lime:water), the mixture was placed into a cubical mold with dimension of 2 cm × 2 cm × 2 cm.

10.6.1 Development of the Bioparticle

Development of the bioparticle involves mixing a binder with other materials and an appropriate volume of water. In this study, the binder used is

TABLE 10.2

Properties of Bioparticle with Different Components Ratio

Bioparticle	Zeolite (g)	Blast Furnace Slag (g)	Water (mL)	Ratio
Bioparticle 1	20	10	10	2:1:1
Bioparticle 2	20	20	10	2:2:1
Bioparticle 3	20	10	20	2:1:2

cement. Zeolite is chosen to be incorporated in the bioparticle since it has proven to be useful in wastewater treatment as cation exchange. The materials were mixed together based on a suitable mix ratio design.

The fundamental constituents of the bioparticle are zeolite, blast furnace slag, and binder. The size of the small zeolite and blast furnace slag ranges from 3 to 5 mm as stated in Table 10.2. This was done by sieving procedure to obtain different sizes of the aggregates. Cement was used as the binder in bioparticle production.

After the mixing with a suitable ratio, the mixture was placed into a mold and air dried for at least 5 days. The determined ratio is the most outstanding and could withstand physical testing such as reliability in acidic and alkaline conditions, high temperature, and placement under flowing water. The mold should be covered with aluminum foil or wrapping plastic to prevent water loss from the mixture. Water retained is needed to strengthen the structure of the bioparticle.

After 5 days, the bioparticles were removed from the mold prior to the curing stage in tap water for at least 7 days. The reason for curing was to develop the biofilm that will attach and grow at the surface of the bioparticle. Indigenous bacteria which already existed in the wastewater itself were used to grow biofilm.

Molds that were used for the development of the bioparticle were in the form of a cube as illustrated in Figures 10.2 and 10.3. Different ratios

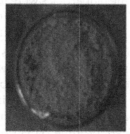

FIGURE 10.2
Composite of bioparticle.

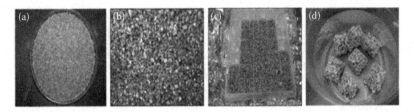

FIGURE 10.3
Bioparticle composite. (a) Pretreated with NaCl solution, (b) blast furnace slag, (c) cubical mold 2 cm × 2 cm × 2 cm, and (d) curing period 2–3 days.

of the compound were used to get particles that are porous. The structure of the bioparticle and its strength during the demolding process was observed.

10.6.2 Bioparticle Reliability Test

The bioparticle reliability test was carried out to select the bioparticles with the best ratio which was suitable in wastewater treatment with high steadfastness. Thus, the bioparticles were tested under certain conditions such as acid conditions, alkaline conditions, and also under water pressure.

1. *Reliability toward Acid Conditions:* Two types of strong acids were used such as sulfuric acid, H_2SO_4 5M (15 mL) and hydrochloric acid, HCL 5M (20 mL). Concentrated sulfuric acid (97%), 137 mL was diluted in 363 mL of distilled water and hydrochloric acid, 207 mL was diluted in 293 mL of distilled water. Bioparticles were then immersed into the diluted sulfuric acid and hydrochloric acid for 3 days. The physical changes of the bioparticles were observed during the first 10 min, 3 h, and finally at day 3 of the test.

2. *Reliability toward Alkaline Conditions:* Natrium hydroxide, NaOH 5M (30 mL) was used for the alkaline reliability test since it was a strong alkaline. It was prepared by dissolving 100 g of NaOH in 500 mL of distilled water. Then, the bioparticles were immersed into the NaOH solution for 3 days. The physical changes of the bioparticles were observed during the first 10 min, 3 h, and finally at day 3 of the test.

3. *Reliability under Water Pressure:* Bioparticles were placed separately in a 100 mL beaker and filled with tap water for 14 days. Results on the physical changes of bioparticles were obtained at the end of day 14 and were recorded. This method is very important because in the treatment all the bioparticles must be filled in the large container for different hydraulic retention time (HRT) treatments. The bioparticles must be in good condition after the reliability under water pressure test.

10.6.3 Designs for Subsurface Constructed Wetland Integrated with Bioparticle

The lab-scale system consists of a storage tank with capacity of 500 L, four cell wetlands, and four effluent collection tanks. All laboratory equipment mentioned above was made of plastic, perspex, and PVC. The system was fabricated in rectangular shape with dimensions of 0.3 m width, 0.3 m length, and 0.5 m depth. Figure 10.4 shows the schematic design of the system and Figure 10.5 shows the lab-scale system that has been set-up.

As shown in Figure 10.5, the lab-scale system of integrated bioparticles and subsurface wetlands was constructed into four sets of configurations known as control, planted bioparticle, and integrated bioparticle in plant. The first compartment in wetland cell was planted with *Heliconia rostrata* while the second compartment is filled with bioparticle without plants. The third compartment was planted with the same plant *Heliconia rostrata* and the last compartment is without plants (control). The wastewater flows into the wetland system by gravity from the storage tank. Figure 10.6 shows the system at the experiment site.

Heliconia rostrata was planted in the cell, saturated with clean water for 2 weeks prior to experimental and monitoring exercises while wastewater was inserted in bioparticle cell for growth of the biofilm.

As shown in Figure 10.7, both the wetland compartment and the control compartment were covered with four layers which are 5 cm depth gravel (3 cm φ), 5 cm depth gravel (1 cm φ), 5 cm depth gravel (0.6 cm φ), and the last layer is soil with 15 cm depth. The reason for designing the compartment

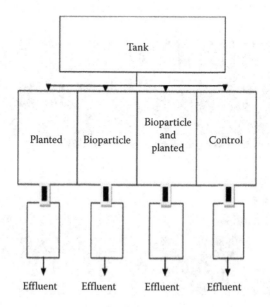

FIGURE 10.4
Schematic design for constructed subsurface wetlands integrated with bioparticle.

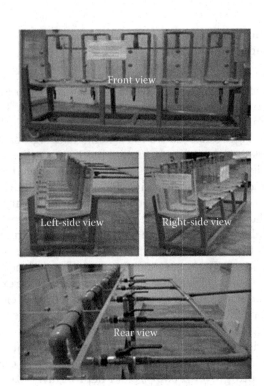

FIGURE 10.5
Lab-scale wetlands cell.

FIGURE 10.6
Wetland system arrangements on site.

FIGURE 10.7
Wetland compartment.

with four layers is to obtain efficiency in filtration. Moreover, the gravel can protect the perforated pipes from being clogged by the media. For the bioparticle compartment, the compartment was filled with bioparticles at 30 cm depth and for the wetlands integrated with bioparticles, the compartment consists of 15 cm depth of bioparticle and 15 cm depth of soil. The water level in the cell will be the same height as the hydraulic pressure principals. The water level in the cell was set to be at the middle level of the soil surface, but for the bioparticle compartment the water level must cover the top of the bioparticle depth to support the biofilm growth. Figure 10.8 shows that new plants were reproduced during the growing period.

10.6.4 Plants

Heliconia rostrata was used as vegetation in the submerged flow (SF) wetland system in this study. This plant was chosen because it was proved to be effective in treating domestic wastewater for tropical climates as stated by researchers from Thailand. The most important functions of this plant are related to their physical effects in the wetlands. The roots provide a huge surface area for attached microbial growth.

FIGURE 10.8
New plants reproduced during the growing period before experiment started.

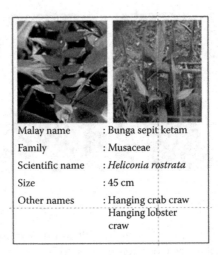

Malay name	: Bunga sepit ketam
Family	: Musaceae
Scientific name	: *Heliconia rostrata*
Size	: 45 cm
Other names	: Hanging crab craw Hanging lobster craw

FIGURE 10.9
Short description of *Heliconia rostrata*.

Heliconia is a fast growing plant, tall-stature, "clonal dominants" that establishes quickly, processes a lot of energy, and considered a suitable plant for treatment of wetlands. Besides that, this plant has its own esthetic value and is mostly used as a landscape plant. Therefore, this plant also increases the esthetics of the wastewater treatment of wetlands. Figure 10.9 shows the basic information and short notes on the physical characteristics of this plant.

Heliconia belongs to the same family as the banana plant. Its height is about 1–2 m. The inflorescence is 30–60 cm long, hanging, and with a pleated look. Reproduction is by young suckers from the rhizomes. This plant is native to Peru.

10.6.5 Components of Wetland Design

1. The most important components in wetlands design are stated in Table 10.3. Hydraulic design flow rate can be calculated using Darcy's Law:

$$Q = k_s AS \tag{10.1}$$

where
  Q = average flow rate through the system, m^3/day
  k_s = hydraulic conductivity of the medium, m^3/m^2-day
  $A = d*W$, cross-sectional area of wetland bed, perpendicular to the direction of flow, m^2
  S = slope of the bed, or hydraulic gradient (as a fraction or decimal)

TABLE 10.3

Experiment Description

	Com. 1	Com. 2	Com. 3	Com. 4
Media	Plant (*Heliconia rostrata*) Soil (15 cm) Gravel—6 mm dia. (5 cm) Gravel—10 mm dia. (5 cm) Gravel—30 mm dia. (5 cm)	Bioparticle (shape: rectangular) Size: 2 cm × 2 cm × 2 cm (15 cm)	Plant (*Heliconia rostrata*) Soil (15 cm) Bioparticle (15 cm)	Soil (30 cm)
Flow rate, Q	0.4 m³/day	0.4 m³/day	0.4 m³/day	0.4 m³/day
Retention time, t	0.03 day	0.03 day	0.03 day	0.03 day
Volume, V	0.011 m³	0.011 m³	0.011 m³	0.011 m³

Note: Com., compartment.

2. Retention Time

Hydraulic retention time can be represented as

$$t = \frac{nLWd}{Q} \tag{10.2}$$

where
t = hydraulic retention time
n = porosity of media, $n = V_v/V$ where (V_v and V are volume of voids and total volume)
L = length of wetlands media, m
W = width of wetlands media, m
d = depth of wetlands media, m
Q = average flow rate through the system, m³/day

The plant was inoculated in the domestic wastewater for at least 2 weeks in order to reach a steady-state condition of bacteria cultivation. After 3 days, the experimental treatment operation was conducted in 5, 7, 9, 11, 13, and 15 days. The tested parameters are as follows: BOD, COD, AN, P, and NO_3. The removals of BOD, COD, AN, P, and NO_3 have demonstrated the performance of wetland plants in treating domestic wastewater.

10.7 Result and Discussion

10.7.1 Compartment Condition and Removal Efficiency

Table 10.4 shows the change conditions in the compartment while Table 10.5 shows the removal efficiency in each compartment in terms of concentration

TABLE 10.4

Physical Condition of Four Compartments before, during, and after Experiment

Compartment	Before	During	After
Wetlands	*Heliconia* plant with space in between	*Heliconia* seen to germinate with new plant	Wetland system decreased domestic wastewater characteristics
Bioparticle	Bioparticles are put into compartment	Biofilm is seen with bioparticles	Domestic wastewater changes color and odor in physical terms
Integrated wetlands + bioparticle	*Heliconia* plant with space in between	*Heliconia* seen to germinate with new plant	Wetland system decreased domestic wastewater characteristics. Domestic wastewater changes color and odor in physical terms
Control	Soil is prepared according to designed layer	Soil is dense because of absorption from domestic wastewater	Wetland system decreased domestic wastewater characteristics

of BOD, COD, AN, P, and NO_3 from the first day until the 15th day. Wetland and bioparticle compartment records maximum removal on the 15th day with concentration of BOD reduced to 5 mg/L, COD reduced to 47 mg/L, AN to 2.4 mg/L, and phosphorus concentration reduced to 2.07 mg/L.

10.7.2 Different Removal Integrated Wetlands and Bioparticle (W+B), Bioparticle, and Control Compartment

Table 10.6 describes the different removal concentrations of integrated wetlands (W+B), bioparticles, and wetland compartments compared to control compartments. Based on Table 10.6, the nutrient groups consist of AN, P, and NO_3. The figure reflects the comparison between organic and nutrient removal before and after treatments and the difference between removals. The influent concentrations for BOD are 220 mg/L (before treatment) and became 66 mg/L in the control compartment, 5 mg/L in the (W+B) compartment, and 132 mg/L after treatment. The maximum BOD removal in (W+B) systems was 27% compared to the control compartment. The result from the analysis showed that the COD removal using (W+B) was higher compared to other compartments. The different removal for (W+B) is 24% compared to bioparticles' 18%.

The analysis of NO_3 records more significant results in the (W+B) compartment. The percentage removal of nitrate using (W+B) is more effective with 4% different removal. The different removal for AN using (W+B) is 11%, more effective. However, for phosphorus removal, the bioparticle compartment

TABLE 10.5

Result and Removal Efficiency Experiment

	Days/Parameter	BOD	COD	Nitrate	AN	P
Compartment 1 (wetland)	Initial	220	498	11.3	17.52	16.03
	3	209	429	1.9	6.72	15.12
	5	150	407	1.6	4.8	14.4
	7	146	393	0.9	4.88	13.86
	9	134	265	0.8	6.96	13.41
	11	130	200	0.3	5.52	12.78
	13	41	93	0.2	6.24	10.62
	15	37	69	0.25	3.84	14.58
Compartment 2 (bioparticle)	Initial	220	498	11.3	17.52	16.03
	3	218	498	497	15.84	16
	5	215	501	437	15.6	15.66
	7	200	494	363	16.2	5.58
	9	191	372	304	16.32	2.88
	11	168	279	219	14.3	2.7
	13	161	186	159	13.2	1.62
	15	132	74	160	13.44	4.59
Compartment 3 (wetland + bioparticle)	Initial	220	498	11.3	17.52	16.03
	3	206	403	9.7	15.52	15.04
	5	185	372	3.2	13.44	10.98
	7	110	306	2.8	12.48	3.06
	9	43	279	1	11.28	2.16
	11	16	138	0.5	6.24	2.07
	13	15	82	0.1	2.88	1.35
	15	5	47	0.7	2.4	2.07
Compartment 4 (control)	Initial	220	498	11.3	17.52	16.03
	3	212	512	3.5	16.8	28.17
	5	227	544	1.7	12.72	23.94
	7	192	556.8	1.1	9.12	23.22
	9	138	372.6	1.1	8.72	17.37
	11	133	345.8	1	7.2	12.87
	13	123	282.9	0.6	7.32	12.78
	15	66	165	0.8	6.36	11.07

COD, chemical oxygen demand; AN, ammoniacal nitrogen; P, phosphorus; BOD, biological oxygen demand: Value obtained is higher than initial * unit mg/L.

TABLE 10.6

Removal Efficiency Compared to Control Compartment

Parameter	Compartment	Influent (mg/L)	Effluent (mg/L)	Removal (%)	Different Removal (%)
BOD	Control	220	66	70	
	Wetland + bioparticle		5	97	+27
	Bioparticle		132	40	−30
	Wetland		37	83	+13
COD	Control	498	165	67	
	Wetland + bioparticle		47	91	+24
	Bioparticle		74	85	+18
	Wetland		69	86	+19
NO$_3$	Control	17.52	0.6	95	
	Wetland + bioparticle		0.1	99	+4
	Bioparticle		159	Over	−
	Wetland		0.2	98	+3
AN	Control	16.03	4.32	75	
	Wetland + bioparticle		2.4	86	+11
	Bioparticle		13.2	26	−49
	Wetland		3.84	78	+3
	Control	11.3	12.78	20	
P	Wetland + bioparticle		10.62	34	+14
	Bioparticle		1.62	90	+70
	Wetland		10.62	33	+13

Notes: COD, chemical oxygen demand; AN, ammoniacal nitrogen; P, phosphorus; BOD, biological oxygen demand.

records 70% effectiveness compared to the (W+B) compartment with just 14% different removal.

10.8 Conclusion

Constructed wetlands have a good potential for wastewater treatment in developing countries due to their simple operation and low implementation costs. The main objective for this study was to determine the effectiveness of wetland and bioparticles (W+B) in treating domestic wastewater and the retention time provided to treat domestic wastewater. The parameters that were analyzed consisted of BOD, COD, AN, nitrate (NO3), and phosphate (PO$_4$) in the constructed wetland with different compartments. The overall study shows positive results where all the parameters tested increased in percentage removal.

The performance of a constructed wetland system in domestic wastewater treatment has generally shown excellent ability in reducing pollutants with a certain removal percentage. Based on the results and observation of this study, the following conclusions can be made:

1. Free water surface constructed wetland system with free-floating wetland plants recorded an excellence performance in treating contaminants such as organic matter (80% of removal efficiency) and nutrients (80% of removal)

2. Wetland plant *Heliconia* has a significant role in the performance of a constructed wetland system

3. Retention time is a significant factor in the performance of constructed wetland system and it is observed that each pollutant has its adequate retention time to achieve the highest removal

10.9 Recommendations

Constructed wetlands have been used worldwide for polishing the effluents of wastewater treatment. Furthermore, there are countries that reuse the effluent from constructed wetland treatment, especially countries experiencing scarcity of water resources. The United States and Australia are two representative countries that reuse the effluent for daily activities such as irrigation and toilet flushing. The application of bioparticles combined with a wetland system will create a good condition to remove contaminants and nutrients from domestic wastewater. Bioparticles and wetland systems are an alternative method to achieve low-cost treatments of domestic wastewater. This application may be suitable for the proposing of an ultimate alternative treatment for sewage system in Malaysia.

Reusing wastewater has been practiced for years. The applications of constructed wetlands systems for domestic wastewater treatment are able to provide water quality suitable for reuse. Further treatment such as chlorination is suggested to ensure the effluent is safe for reuse without introducing any side effects or threats to human beings. Suggested domestic wastewater-reuse activities such as irrigation, toilet flushing, and vehicle washing do not require a high water quality standard. Most importantly, the return of cleaned water to the ecosystem helps increase aquifer recharge and river flow, allowing for more natural cleaning through the river system.

References

Ayaz, S.C. and Akca, L. 2001. Treatment of wastewater by natural systems. *Environment International*. 26:189–195.

Chingomb, P., Saha, B., and Wakeman, R.J. 2005. Surface modification and character-ization of a coal-based activated carbon. *Carbon*. 43:3132–3143.

Crini, G. 2006. Non-conventional low-cost adsorbents for dye removal: A review. *Bioresource Technology*. 97:1061–1085.

Faria, P.C.C., Orfao, J.J.M. and Pereira, M.F.R. 2004. Adsorption of anionic and cationic dyes on activated carbons with different surface chemistries. *Water Research*. 38:2043–2052.

Haidiahvita, B.S. 2005. Pembangunan biopartikel sebagai media tapisan dalam penapis-bio untuk rawatan air sisa tercemar. Tesis B.Sc. UniversitiTeknologi Malaysia.

Hayashi, J., Kazehaya, A., Muroyama, K., and Watkinson, A. P. 2000. Preparation of activated carbon from lignin by chemical activation. *Carbon*. 38:1873–1878.

Lim, C.K., Aris, A., Neoh, C.H., Lam, C.Y., Majid, Z.A., and Ibrahim, Z. 2014. Evaluation of macrocomposite based sequencing batch biofilm reactor (MC-SBBR) for decolorization and biodegradation of azo dye Acid Orange 7. *International Biodeterioration and Biodegradation*, 87:9–17.

Nurfarahain, B.M.R. 2008. Application of bioparticle for styrene rich wastewater treatment. *Bachelor of Degree*, Universiti Teknologi Malaysia, Skudai.

Wang, S.B., Peng, Y.L. 2010. Natural zeolites as effective adsorbents in water and wastewater treatment. *Chemical Engineering Journal*. 156:11–24.

Bibliography

Bell, J.G. 1998. Vegetation and water quality monitoring of a constructed wetland for treatment of urban stormwater runoff. Master Degree Thesis. The University of Calgary.

Bingham, D.R. 1994. Wetlands for stormwater treatment. In: Kent, D.M. (ed.), *Applied Wetlands Science and Technology*. Florida: Lewis Publishers, pp. 243–262.

Brix, H. 1993. Wastewater treatment in constructed wetlands: System design, removal processes, and treatment performance. In: Moshiri, G.A. (ed.), *Constructed Wetlands for Water Quality Improvement*. Boca Raton: CRC Press Inc., pp. 9–22.

Corbitt, R.A. 1998. *Standard Handbook of Environmental Engineering*. 2nd ed. New York, NY: McGraw-Hill.

Cronk, J.K. and Fennessy, M.S. 2001. *Wetland Plants: Biology and Ecology*. Florida: Lewis Publishers.

Des, B., Prakash, S., Reddy, P.S.R., and Misra, V.N. 2007. An overview of utilization of slag and sludge from steel industries. *Resources, Conservation and Recycling* 50:40–57.

Erdam, E., Karapinar, N., and Donat, R. 2004. The removal of heavy metal cations by natural zeolites. *Journal of Colloid and Interface Science* 280:309–314.

Greenway, M. and Simpson, J.S. 1996. Artificial wetlands for wastewater treatment, water reuse and wildlife in Queensland, Australia. *Water Science and Technology*. 33(10–11):221–229.

Hammer, D.A. 1992. *Creating Freshwater Wetlands*. Boca Raton, FL: Lewis Publishers, 1991.

Juwakar, A.S., Oke, B., Juwarkar, A., and Patnaik, S.M. 1995. Domestic wastewater treatment through constructed wetland in India. *Water Science Technology*, 32(3):291–294.

Kadlec, R.H. 1995. Overview: Surface flow constructed wetlands. *Water Science Technology*, 32(3):1–12.

Kavitha, A. and Ganesan, P. 2007. Application of bioparticle for the treatment of petrochemical wastewater with and without biofilm. Tesis B.Sc. UniversitiTeknologi Malaysia.

Martin, C.D. and Johnson, K.D. 1994. The use of extended aeration and in-series surface flow wetlands for landfill leachate treatment. *Water Science Technology*, 32(3):119–128.

Mashauri, D.A., Mulungu, D.M.M., and Abdulhussein, B.S. 2000. Constructed wetland at the University of Dar Es Salaam. *Water Research*. 34(4):1135–1144.

Metcalf and Eddy, 1999. *Wastewater Engineering*. 4th ed. McGraw-Hill, New York.

Mohamad Lazim, M.A.B., Neoh, C.H., Lim, C.K., Chong, C.S., Ibrahim, Z. 2014. Biofilm-coated macrocomposites for the treatment of high strength agricultural wastewater. *Desalination and Water Treatment*, 57(8):3424–3429. doi: 10.1080/19443994.2014.989910.

Nadirah, I. 2005. Aplikasi penapis biologi dan fitoremediasi dalam merawat air sisa tercemar. Tesis B.Sc. Universiti Teknologi Malaysia.

Polprasert, C., Rajput, V.S., Donaldson, D. (reviewer), and Viraraghavan, T. (reviewer) 1982. *Septic Tank and Septic System*. Bangkok, Thailand: Environmental Sanitation Information Center.

Rodrigues, V. 1997. Constructed wetlands for tertiary sewage treatment and wildlife habitat in Nova Scotia. Master Degree Thesis. Dalhousie University.

Scodari, P.F. 1990. *Wetlands Protection: The Role of Economics*. Washington, DC: Environmental Law Institute.

Senzia, M.A., Mashauri, D.A., and Mayo, A.W. 2003. Suitability of constructed wetlands and waste stabilization ponds in wastewater treatment: Nitrogen transformation and removal. *Physics and Chemistry of the Earth*. 28:1117–1124.

Sewerage Service Department, Ministry of Housing and Local Government. 2000. *Guidelines for Developers (Septic Tank)*. Malaysia, ISBN 983-2190-03-7.

Shutes, R.B.E. 2001. Artificial wetlands and water quality improvement. *Environment International*. 26:441–447.

Stottmeister, U., Wiepner, A., Kuschk, P., Kappelmeyer, U., Kastner, M., Bederski, O., Muller, R.A., and Moormann, H. 2003. Effects of plants and microorganisms in constructed wetlands for wastewater treatment. *Biotechnology Advances*. 22:93–117.

Sundaravadivel, M. and Vigneswaran, S. 2001. Constructed wetland for wastewater treatment. *Environmental Science and Technology*. 31(4):351–409.

Tanner, C.C. 1999. Plants as ecosystem engineers in subsurface flow treatment wetlands. *Journal of Environmental Quality*. 27(2):448–458.

Wen, D., Ho, Y.S., and Tang, X. 2006. Comparative sorption kinetic studies of ammonium onto zeolite. *Journal of Hazardous Materials*. 133:252–256.

Index

Printed in the United States
by Baker & Taylor Publisher Services